BOOKS MAY BE RENEWED

D1287344

WOMEN IN MATHEMATICS

Scaling the Heights

Cover design by Barbieri & Green

©1997 by the Mathematical Association of America

ISBN 0-88385-156-3
Library of Congress Catalog Number 97-74346

Printed in the United States of America

Current Printing
10 9 8 7 6 5 4 3 2 1

33.95

WOMEN IN MATHEMATICS

Scaling the Heights

Deborah Nolan, Editor
University of California, Berkeley

Published by

THE MATHEMATICAL ASSOCIATION OF AMERICA

MAA Notes Series

The MAA Notes Series, started in 1982, addresses a broad range of topics and themes of interest to all who are involved with undergraduate mathematics. The volumes in this series are readable, informative, and useful, and help the mathematical community keep up with developments of importance to mathematics.

MAA Notes

11. Keys to Improved Instruction by Teaching Assistants and Part-Time Instructors, *Committee on Teaching Assistants and Part-Time Instructors, Bettye Anne Case,* Editor.

13. Reshaping College Mathematics, *Committee on the Undergraduate Program in Mathematics, Lynn A. Steen,* Editor.

14. Mathematical Writing, by *Donald E. Knuth, Tracy Larrabee, and Paul M. Roberts.*

16. Using Writing to Teach Mathematics, *Andrew Sterrett,* Editor.

17. Priming the Calculus Pump: Innovations and Resources, *Committee on Calculus Reform and the First Two Years,* a subcomittee of the Committee on the Undergraduate Program in Mathematics, *Thomas W. Tucker,* Editor.

18. Models for Undergraduate Research in Mathematics, *Lester Senechal,* Editor.

19. Visualization in Teaching and Learning Mathematics, *Committee on Computers in Mathematics Education, Steve Cunningham and Walter S. Zimmermann,* Editors.

20. The Laboratory Approach to Teaching Calculus, *L. Carl Leinbach et al.,* Editors.

21. Perspectives on Contemporary Statistics, *David C. Hoaglin and David S. Moore,* Editors.

22. Heeding the Call for Change: Suggestions for Curricular Action, *Lynn A. Steen,* Editor.

24. Symbolic Computation in Undergraduate Mathematics Education, *Zaven A. Karian,* Editor.

25. The Concept of Function: Aspects of Epistemology and Pedagogy, *Guershon Harel and Ed Dubinsky,* Editors.

26. Statistics for the Twenty-First Century, *Florence and Sheldon Gordon,* Editors.

27. Resources for Calculus Collection, Volume 1: Learning by Discovery: A Lab Manual for Calculus, *Anita E. Solow,* Editor.

28. Resources for Calculus Collection, Volume 2: Calculus Problems for a New Century, *Robert Fraga,* Editor.

29. Resources for Calculus Collection, Volume 3: Applications of Calculus, *Philip Straffin,* Editor.

30. Resources for Calculus Collection, Volume 4: Problems for Student Investigation, *Michael B. Jackson and John R. Ramsay,* Editors.

31. Resources for Calculus Collection, Volume 5: Readings for Calculus, *Underwood Dudley,* Editor.

32. Essays in Humanistic Mathematics, *Alvin White,* Editor.

33. Research Issues in Undergraduate Mathematics Learning: Preliminary Analyses and Results, *James J. Kaput and Ed Dubinsky,* Editors.

34. In Eves' Circles, *Joby Milo Anthony,* Editor.

35. You're the Professor, What Next? Ideas and Resources for Preparing College Teachers, *The Committee on Preparation for College Teaching, Bettye Anne Case,* Editor.

These volumes can be ordered from:
MAA Service Center
P.O. Box 91112
Washington, DC 20090-1112
800-331-1MAA   FAX: 301-206-9789

To

Leon Henkin

and

Barbara Waterman

for their inspiration and encouragement.

Preface

On July 14, 1994, a group of mathematicians from around the country gathered in Berkeley for three days to discuss ways to increase the representation of women in PhD programs in the mathematical sciences. The conference, entitled "Women in Mathematics: Scaling the Heights and Beyond," was organized by the directors of the Mills College Summer Mathematics Institute (SMI)—Ani Adhikari, Lenore Blum, Steven Givant, Leon Henkin and myself—and funded by the National Science Foundation.

The primary goal of the conference was to broaden the impact of the SMI, a summer program designed to encourage and prepare talented undergraduate women to attend graduate school in the mathematical sciences. We wanted to assist mathematicians from across the US in developing their own programs to advance women in mathematics.

In response to our conference announcements, mathematicians from six institutions came forward with ideas for programs, and many others expressed an interest in helping develop such programs. Given the strong response to our announcement, we enlarged the scope of the conference. We invited representatives from mathematical professional societies, organizers of existing successful math programs, and other mathematicians who were interested in these issues. Altogether, more than 40 mathematicians participated in the conference.

We audio-recorded the speeches and discussions at the conference in order to publish a conference report. I took on this task, and one year later, with the speeches transcribed and edited, I decided to enlarge the scope of the report. My reason for doing this was that two important aspects of the conference were missing from the report. The conference participants had the benefit of meeting the undergraduate women who were part of the SMI that summer. They heard these women speak on topics such as their math background, their future career plans, and why they chose to attend the summer program. The conference participants also had the opportunity to observe the SMI faculty in action. They saw, firsthand, the teaching strategies and course materials the faculty had developed in order to create an exciting and challenging environment in which to learn mathematics. I thought that adding these elements to the book would increase its interest and usefulness to the mathematics community.

In its present form this collection of articles contains not only speeches from the conference, but also examples of math courses from the SMI, descriptions of successful summer math programs, and a survey of undergraduate math majors.

The articles are organized into three distinct parts. The first contains: Lenore Blum's introductory speech on the progress of women in mathematics in the past 20 years; Deborah Haimo's keynote speech on how to achieve excellence in mathematics by making its study inclusive for all those with interest and ability; and Carol Wood's speech describing some of the programs, present and planned, of the Association for Women in Mathematics. These speeches are complemented by an article that includes the voices of today's undergraduate math majors. They express thoughts on their major, on their academic performance, and on their future plans in mathematics. The gender comparisons are striking. The material for this article is compiled from a quantitative survey of the math majors at the University of California at Berkeley and from a review of the applications for the Mills SMI.

In the second section, several SMI faculty provide answers to the question:

How do we design upper-level courses to attract women and help them excel?

The faculty were invited to describe their SMI courses. The main strength of the SMI lies in its exceptional courses. At the conference, it was generally agreed that these courses were crucial to generating student interest in studying advanced math. In their articles, the faculty share detailed outlines of course material, lists of challenging problems, sample research projects, valuable references, advice on how to incorporate their techniques into the traditional classroom, and student reactions to the course. It is a great resource. The material here is not gendered. It is our hope that these articles will be used to design exciting upper level courses for all undergraduate math majors.

The last section contains descriptions of several summer programs. At the conference, six new programs were proposed and discussed, and several others were offered as models for designing these new programs. One month after the conference, a joint proposal for funding the Mills SMI and these six new programs was submitted to the National Science Foundation and the National Security Agency. Three were funded, and are still in operation today. They are the Carleton and St. Olaf Colleges' Summer Mathematics Program, the Summer Program for Women in Mathematics at the George Washington University, and the Mills College SMI, which has since moved to UC Berkeley and is now called the Summer Institute for the Mathematical Sciences (SIMS). These three programs are described in separate articles in the third section. In addition, other articles describe the summer programs at Mt. Holyoke College, the University of Michigan, the National Security Agency, and Spelman and Bryn Mawr Colleges.

Finally, I would like to acknowledge all those who have contributed to this collection of articles. The conference was a great success due to: the enthusiastic participation of the speakers, discussants, and attendees; the organizational assistance provided by Kathy Guarnieri; and the support of the National Science Foundation. (The conference participants are named in the Appendix.) Jane Muirhead and Chris Bush had the painstaking job of transcribing the speeches from the conference. My undergraduate, upper-level statistics class—STAT 135, Fall 1997—helped conduct the survey of UC Berkeley math majors. Lubna Chunawala had the arduous task of preparing the initial draft of this document. Faye Yeager assisted in preparing subsequent drafts. I am most indebted to the SMI instructors who generously gave their time to write descriptions of their courses in order that they may be used by faculty everywhere.

Deborah Nolan
Statistics Department
University of California, Berkeley
March, 1997

Contents

On *Women* *in* *Mathematics*

Women in Mathematics: Scaling the Heights and Beyond [1]

Lenore Blum, Deputy Director

Mathematical Sciences Research Institute (MSRI), Berkeley, CA

To help set the stage for our discussions, I'd like to take this opportunity to address briefly three points:

- A 20+ year perspective
- Why a program for women?
- Where do we go from here?

A look back

In 1971, the Association for Women in Mathematics (AWM) was founded. In January of that year, at the national meeting of the American Mathematical Society (AMS) in Atlantic City, Mary Gray led a sit-in protesting closed Council sessions that effectively excluded the participation of women from professional activities of the Society. Shortly afterwards, she circulated a Newsletter and solicited members for a new organization. Later that spring, in conjunction with protests on the Berkeley campus about the paucity of women in regular faculty positions, I organized the first colloquium on women in mathematics in the Berkeley mathematics department. Soon the east coast and west coast forces joined, and the Association for Women in Mathematics was clearly established. From the start, the AWM has worked to increase the participation, and improve the position, of women in mathematics (Blum, 1987, 1991). Carol Wood, past president of the AWM, will speak later on some of the past and current activities of the Association.

What was the education of girls and women in mathematics like in the early seventies? In 1973, I went to Mills College, a small liberal arts college for women, as a visiting assistant professor. I was surprised to find there was no mathematics department at Mills (mathematics was part of chemistry and physics).

So, I wrote a position paper on how (and why) to design a mathematics and computer science department, which I circulated around campus. A major outcome was that the president of Mills decided to start the new department, and appointed me as its head. In those early years, we focused on effective access into the mathematics curriculum at the entry level. One key aspect of the Mills program was to create an alternative to the traditional college algebra course for women who had not had enough mathematics in high school. College algebra was a dead end for most of these students; they never went farther in mathematics. So instead, we developed a streamlined precalculus/calculus sequence that would enable our students to enter the regular mathematics curriculum quickly. One of the first things I did when I was at Mills College was to hire Steve Givant to help develop and expand that program (Blum and Givant, 1980).

This situation was not peculiar to Mills. In 1973, Lucy Sells, a sociology graduate student at Berkeley, surveyed entering students on that UC campus. She found some striking differences between the preparation of women and men for the entry-level calculus sequence, the prerequisite for all of future course work in mathematics, science and engineering. Although her study was not systematic, it was important since it pinpointed high school mathematics education as a "critical filter" for careers in mathematics and science. About the same time, Julia Sherman and Elizabeth Fennema, two mathematics educators in Madison, were studying girls in mathematics in elementary and high schools. They found that there was a critical period in the third and fourth years of high school when mathematics becomes an elective. At this stage, girls were counseled out of mathematics by their parents and their teachers who did not seem to see the value of four years of high school mathematics for their daughters' futures. Essentially, these girls would be told: Why spoil a perfectly good high school record with a possible B or C in mathematics? (Boys, on the other hand, were generally encouraged to stick with it.)

In 1973, Nancy Krienberg found the situation similar for younger girls. She had started after-school mathematics and science classes for grade school children at the Lawrence Hall of Science (LHS) in Berkeley, but noticed that there were very few girls in the programs. This led her to develop Math for Girls, an after-school mathematics program specifically for elementary school girls.

[1]Introductory Remarks, Mills Summer Mathematics Institute Conference on Programs for Women in Mathematics, July 14–16, 1994, Berkeley.

3

In the summer of 1975, a number of us met at the LHS to compare experiences about successful educational strategies. We found that, independent of age level or institution, successful programs had fundamental components in common. Namely, they provided: information to girls and women about the importance of mathematics for their futures; effective access into the mathematics curriculum and positive hands-on experiences doing mathematics; and role models and mentors. Out of this meeting came the Math/Science Network and its subsequent activities; these include the Expanding Your Horizons conferences for junior and high school girls and the EQUALS teacher training programs, both of which have had considerable national and international impact.

In the seventies, very few mathematicians were involved in programs for increasing women in mathematics. The impetus came very often from women's centers and women's groups. One notable exception, coming from the mathematics community, was John Ernest's influential booklet and article, "Mathematics and Sex" (1976). More typical was a feature article in *Ms.* magazine (September, 1976) on "Math Anxiety" by social scientist Sheila Tobias which helped put the idea in the public domain.

Although women's groups identified the problem of the dearth of women in mathematics early on, they generally didn't know how to design effective programs to combat the problem, because, for the most part, they were not mathematicians or scientists. Often the result was the development of math anxiety clinics where people talked about negative experiences in their early mathematics education. The clinics might help people feel better, but they did not provide significant tools to do mathematics.

Many mathematicians, myself included, went to the opposite extreme, criticizing the words "math anxiety," preferring instead to use "math avoidance." We were trying to say very strongly that the best way to get people to overcome their avoidance of mathematics was to provide successful experiences doing mathematics. And also, there was another deeper, subtler point—tolerating some level of anxiety in doing mathematics is probably a good thing for mathematical creativity.

So there was quite a spectrum of programs. In those years, I traveled around the country a lot, acting as liaison between mathematics departments and women's centers, helping these groups talk together and jointly develop programs. Over the years, many

very substantial and very good programs were developed across the country.

Jumping ahead

Now, let us jump ahead about 15 years or so to 1990, to the time we submitted our proposal for the Mills Summer Mathematics Institute (SMI). The Mills SMI was to be a program for the most talented women undergraduates in mathematics in the country. This was quite a switch—from programs for entry-level students who might not have thought of going into mathematics to a program for students who have already shown talent in advanced undergraduate mathematics. It was an extremely controversial proposal. We received a lot of criticism of the sort: these women are talented, they are already majoring in mathematics in college, they are doing well already, why do we need to pay any special attention to them?

The fact is, many very talented women who have taken a lot of mathematics courses decide not to continue their studies at the graduate level in mathematics or the sciences. They have great ambivalence about going ahead. We are losing the cream of the women in this country who could go on to study mathematics. A short six-week intervention can have a huge impact on these students. In contrast to the entry-level programs, where it is crucial to sustain a follow up, a single program here can have a lasting effect. That is one really delightful thing about working with these students.

Also asked was, why have a special program for women? Is it because women have to learn mathematics in a special way?

To the contrary, it is to create an environment where women have a chance to do and learn mathematics in ways that most successful male mathematicians take for granted. Mathematics is a social and communal endeavor; mathematicians enjoy meeting with each other in their field to work, talk, and play together. Many budding male mathematicians enjoy having friends in high school and college with whom they can jointly work on mathematics problems. Much college mathematics is done in dormitory dining rooms and mathematics department common rooms. There is a natural flow between the personal and professional. For example, it is not uncommon for MSRI visitors to spend a day sailing on the bay with their colleagues, thus promoting professional bonding. These experiences are rarely part of women mathematicians' experiences. The opposite is more often the rule—women in mathematics are very often isolated. Indeed, given this isolation, it's a wonder that women can compete at all in this realm. An important function of the Mills SMI is to capture some of these crucial experiences, and to have the students recognize the value of cultivating such interactions (Forman, 1992).

What is the success of such programs? The Mills program has been in operation since 1991. Each year we have about 20 students. Altogether, from 1991 to 1996, 137 students have participated in the Mills Summer Mathematics Institute. At the start of the program, we find considerable ambivalence amongst the students about their future plans for graduate work in mathematics, but by the end of the six weeks, we often see a complete turn-around in terms of these attitudes. You can see this turn-around in a lot of the exit comments of the students. Steve Givant will address this in his talk. We see very important networks forming and students planning to meet again at mathematics meetings. We start seeing these students popping up at other mathematics programs. For example, two of the teaching assistants this summer are former SMI participants; others go on to REUs (Research Experience for Undergraduates), or to the Budapest program. Several former students attended the Women in Algebraic Geometry workshop held last spring at MSRI (Blum, 1993). Many go on to graduate school.

Deb Nolan provided the following figures for the program. Of the 24 participants in the 1991 program, one has received her PhD in math, and three have received MAs, two in statistics and one in operations research. Fourteen are currently enrolled in PhD programs in the mathematical sciences, and the remaining six have careers as math teachers, actuaries, consultants, and software developers. Thirteen of the 25 students from the 1992 program are currently in PhD programs in math and statistics. Eight of the remaining 12 are working in math-related areas (two have Master's degrees). Eighteen students participated in the 1993 program, 11 are currently in math PhD programs, and one is in a Master's program in education. The remaining six are working in finance, software development, and math education at the high school level.

Something not measured by these numbers is the subtle yet significant experiences these students acquire that help them make a successful transition to graduate school and professional life. My sense is that these students develop better professional attitudes and survival skills for graduate school and beyond. They have already had the experience of interacting closely with faculty and graduate students. They have become part of networks of colleagues and peers (e-mail has played an important role here). Graduate school is not scary for them. They know that if they have a problem, they can go up to their professors and talk about it. Most importantly, they seem to have gained more self-confidence about doing mathematics, certainly a key component for success.

Going forward

So where do we go from here? The focus of these new programs is for women in mathematics to scale the heights, and beyond. We want these students to have the opportunity to become successful professional mathematicians. I do not think there is a single way, a single model. I think there are many possibilities, and there are many good ideas. What is really important is for designers and organizers of new programs to do what is most natural and appropriate for them and for their institutions. That will be what works best.

References

Blum, L. (1987). Women in Mathematics: An international perspective eight years later. *The Mathematical Intelligencer* **9**: 28–32.

Blum, L. (1991). A brief history of the Association for Women in Mathematics: The Presidents' perspectives (1971-1991). *Notices of the American Mathematical Society* **38**: 738–754.

Blum, L. (1993). Women in algebraic geometry workshop at MSRI. *Notices of the American Mathematical Society* **40**: 960–962.

Blum, L. and Givant, S. (1980). Increasing the participation of women in math-based fields: A collegiate program. *The American Mathematical Monthly* **87**: 785–793.

Ernest, J. (1976). Mathematics and sex. *The American Mathematical Monthly* **83**: 595–614.

Forman, S. (1992). The Mills women. *Focus (The Newsletter of the MAA)* **12**: 16–17.

Excellence in Mathematics[1]

Deborah Tepper Haimo

*Visiting Scholar, University of California,
San Diego*

What do we mean by excellence in mathematics? Whom are we talking about, and at what level?

Most of us would undoubtedly associate the designation "excellence in mathematics" with the names of some of the great mathematicians whose works revolutionized the field. It is they whose contributions were often so spectacular that we still marvel at the depth of their insight.

The structures of societies and their cultures necessarily led to distinguished mathematicians' being nearly always men, men whose achievements went such a long way toward shaping the discipline that their impact continues to this day. Quite naturally, they set the standards for proficiency in the field. At the same time, it is hardly surprising that they transmitted to it a male cast in the sense that gender became an unstated and implicit characteristic for becoming a mathematician.

The emergence of a few extraordinary minds depends on having many others working in the discipline at various levels of excellence. There must be many in the field to create an environment of knowledgeable people who can appreciate, identify, and develop excellence. Thus, for mathematics to flourish, it seems reasonable to develop the talents of potential mathematicians whoever they may be, and not to discourage some for reasons unrelated to their abilities. For too long, we have neglected a major segment of our population – women. We have failed to encourage them to see mathematics as an attractive, challenging, and fulfilling career.

Up until the past two or so decades, whereas males who indicated an interest and proficiency in mathematics were taken seriously and encouraged to continue in the field, women with like bent were not considered creditable and were largely ignored. It is no secret that students who undertook to study mathematics at nearly every level were for the most part male, their proportion increasing rapidly as the subject became more advanced. At the faculty stage in mathematics, at nearly all research universities and many other educational institutions, members were almost always male, and this continues in many departments to this day.

Indeed, the image of a mathematician, prevalent throughout the world, has been that of a man. It also has become a generally accepted fact that creativity for a mathematician must be manifest at a young age. This has led to a very difficult problem for women who then feel that, during critical years in their lives, they must face the alternatives of having families or concentrating on their professional growth. The result has been that most women who had any interest in mathematics tended to believe what they heard, doubted that they could have mathematical ability, and turned their energies to other pursuits where they felt more welcomed.

The growing awareness of this situation in the United States has resulted in some change in attitudes and policies. There is more and more acceptance of women who show an interest in, and the inclination to study, mathematics. Far more than that, many male mathematicians, along with their female counterparts, are deeply involved in trying to bring about a more balanced situation by providing added support and encouragement for women who are considering mathematical careers.

I thus want to emphasize what most of us already know: excellence may exist anywhere, and the potential for greatness in mathematics is not restricted to one gender. Indeed, excellence can certainly be found in the achievements of those undergraduate female students with the capacity to do advanced graduate work, who succeed in attaining strong doctorates from well recognized, demanding departments. Their accomplishments are often all the more remarkable in that they must overcome society's general disapproval of their activities. This disapproval is transmitted by the prevalent culture where even men who are mathematicians are often regarded as boring, anti-social "nerds," though harmless, and perhaps totally unworldly, yet generally lovable. How then can women mathematicians seem – virtual freaks! Is there a word to describe a female "nerd"?

This explicit social discouragement, unfortunately, does not stop at the gates of academe. Even among those who understand and value mathematics, there is often expressed a bias against women, subtle or not so subtle, by their own instructors, their peers, and, in some cases, even members of their own families. It is thus doubly important that we recognize

---

[1] Keynote speech given at the Mills Summer Mathematics Institute Conference on Programs for Women in Mathematics, July 14–16, 1994, Berkeley.

their accomplishments and provide them with the opportunities to continue in mathematics until they establish themselves in their fields.

That already there have been some dramatic changes in mores is clear to many of us who can still remember different times. When I was college bound, for example, and MIT was one of the schools to which I was admitted, friends of my parents, in line with the conventional views widely held in those days, urged that I go to MIT, arguing that it was an ideal place for me to find a suitable husband.

Fortunately, my parents were more enlightened and did not try to dissuade me from my choice of Radcliffe. There was no escape even there, however. Our first English assignment was to write an account explaining our choice of school, the instructor commenting that he realized that our primary motivation was to find a husband at Harvard. Women were presumed to be interested only in marriage and to come to college merely to make what would be regarded as a "good catch"! As things turned out, I did marry a Harvard graduate student in mathematics, and I also obtained a good education, my goal when I initially selected Radcliffe.

Fortunately for women and for mathematics, the EXPLICIT discouragement of women from striving for professional achievement has almost ended in our educational institutions. I am sure I don't have to convince all of you that major problems still exist. We have a long, difficult, indeed awesome task ahead of us. The overt, flagrant actions of the past are now covert, very subtle, and much harder to overcome. Our challenge is to bring about significant and lasting change in the culture, so that all with an interest in mathematics can pursue its study in an inviting environment, and be given encouragement to advance to the full extent of their abilities.

When she heard the title of my talk, one of my friends asked why I had not called it "Changing the Culture." My response was that the two are interrelated, since I believe that the effect of changing the culture will result in making the study of mathematics inclusive of all those with an interest and ability in the subject, but excellence in mathematics will be based on a far broader and more representative population. At present, we are concentrating our efforts on achieving excellence with our emerging female mathematicians by providing them with some extra support.

We had noted earlier that mostly men achieved greatness in mathematics. Nonetheless, others not satisfying the standard model of a young man of means from a well-developed nation serve as notable counterexamples. We all have our favorites who surmounted great odds to reach the height of their profession. I doubt that anyone would challenge, for example, my inclusion of Emmy Noether, the distinguished algebraist, a woman; Srinivasa Ramanujan, the remarkable number theorist from a small Indian village, certainly not wealthy; and Karl Weierstrass, the noted analyst whose contributions continued into his advanced years when he was no longer young.

These, of course, are regarded by some as anomalies, but even the giants whom we described earlier are anomalies that emerge from a large number of those working in the field. By failing to bring out all possible talent, we are missing a great opportunity for a tremendous pool of potential excellence.

A discipline like mathematics that demands objective reasoning, with results subjected to most careful scrutiny before being validated, is blind to any issues of gender, race, age, or economic status. Thus, it is unfortunate that not all mathematicians accept the notion that excellence in mathematics may be attainable by anyone with sufficient ability.

Reports in recent years have stressed that the decrease in the numbers of US citizens who are males and who are continuing their advanced studies toward doctorates in mathematics endangers our preeminent position in the world, particularly as we are in an ever increasing technological era. Further, many studies have shown that although there appear to be differences in approach to the learning of mathematics between some males and females, in a welcoming climate, gender differences generally disappear in the final results.

Despite these observations, there are still some who are convinced that, by including underrepresented constituencies among those who study at advanced levels, we can only reduce standards. This perception is so pervasive that it lurks in the thinking of otherwise reasonable and well-intentioned people. Indeed, many of us women subconsciously accept such perceptions, often without any realization that we are harboring such views.

Our current situation undoubtedly has hastened us to the awareness of the fallacy of concluding that only men can do mathematics. To maintain a leadership role in world mathematics, it is vital that we educate a larger number of our citizens to full recognition that underrepresented groups provide a yet fully untapped resource. Rather than lower standards, it is imperative that we raise expectations!

As we have found in various special programs, talented women not only exist, but can become as fascinated by advanced mathematics as their male counterparts, all the more so if given comparable encouragement. Mathematics is an exciting field, and that excitement, as well as its inherent beauty, must be transmitted to students.

It's unfortunate that the general reaction to providing everyone with the opportunity to learn mathematics has, for the most part, not addressed the problem of those particularly gifted in the area. They, after all, are the future leaders in the field. As a recent member of a visiting committee to one of our most prestigious mathematics departments, I heard the plaintive concern of some of the brightest students. A succinct remark that summed up their frustration was voiced by one who insisted, "Just because we are regarded by everyone as capable does not mean we don't need guidance! We also flounder and need direction!"

Many of us became engaged in mathematics when we were introduced to proofs in geometry. I remember my own thrill when I discovered the validity of a direct proof, which I had thought failed as an alternative to the standard indirect proof because it did not hold when the geometric figures were positioned in a given way. I had forgotten the existence of an axiom that we had hitherto never used, allowing the repositioning of figures without changing their properties.

Proofs in mathematics are vital. They are the essence of the subject, and I regret to see that their importance is being downplayed in the current effort to entice a greater number of general students to study more mathematics. Since the abstract nature of mathematics is a feature that attracted most of us, it seems reasonable to believe that it would appeal also to those of our students who are drawn to the more challenging aspects of the field. Indeed, the remarkably great power of mathematics for broad and diverse application is derived from its abstract nature, and some students might very well prefer to follow their interests in disciplines that are mathematically based and where applications are paramount.

In my own case, I entered college with a pronounced love of mathematics, but no idea of what I could do with such a major when I graduated. I knew that I was temperamentally unsuited to teach at the high school level. That seemed basically the only possibility at the time, and quite unattractive to me, if for no other reason than the fact that female teachers could not keep their posts in Boston public schools if they were to marry, and those were the only schools I knew.

While I considered physics a more reasonable alternative from an employment point of view, an early experiment quickly dispelled that notion. In introductory physics, we did an experiment where iron filings were scattered on the lab table; a magnet was then introduced; and we were to observe and record the positioning of the filings about the poles. Everyone performed the experiment readily except for me! My iron filings would not act as expected. When my repeated efforts failed time and again, the entire lab became intrigued. Everyone, including the laboratory instructor, sought to explain the situation. One observant student finally resolved the problem by noting that right under my lab space was a drawer full of – you guessed it – magnets!

That episode had a great impact on me. It convinced me that, regardless of the consequences of seeking a career in a field with virtually no employment opportunities, mathematics had a relevance to me that no other discipline held. Mathematical results were based on assumptions over which I had some control, as, for example, the earlier geometry proof. In physics, on the other hand, I could reach a result based on some unseen "drawer of magnets" that could completely distort my conclusions!

In this day and age, women can find many opportunities with a bachelor's degree in mathematics, and are not necessarily restricted by irrelevant factors. Indeed, at that level, substantial progress has been made, and women now form a significant proportion of undergraduate mathematics concentrators at most institutions.

The major problem exists at the next level when the dropout rate between those who terminate their studies when they graduate and those who continue on to graduate school is far higher for women than for men. Many reasons have been advanced for this difference. It is certainly true that there are very few role models for women, with many departments having none at all in tenured positions. Further, a number of findings have established that many men in research universities have a supportive environment and subtle encouragement, factors conducive to advancement in mathematics. On the other hand, others, and women in particular, are left to endure feelings of isolation. They often find their environment not only uninviting, but they perceive their retention as being regarded as unimportant, factors that appear to affect women to an even greater degree than men.

There are no clear guidelines as yet that are known to be effective and long-lasting that can be followed to change the culture. There are no simple or quick solutions to the problems we face, but they must be considered and dealt with, as we cannot afford any longer to ignore their existence. Fortunately, there are a number of programs throughout the country now addressing some of these problems. Though slow, the number of women who are beginning to realize their mathematical potential is rising. As more and more doctorates are granted to more and more women in our foremost universities, and they are given opportunities to work with leaders in the field as equals, our expectation is that special programs will no longer be needed.

Meanwhile, it is incumbent upon us to eradicate problems wherever they appear. To this end, it is important that women students be challenged in a friendly learning climate and that they be provided with ample occasions for meaningful interaction, not only amongst themselves, but with their male counterparts, and with a broad range of as many people in mathematics at all levels as is possible.

Here let me emphasize that I believe it important, especially in a female environment, to be sure that women are exposed to male mathematics students so that both learn early how their counterparts deal with similar problems and how they think. The world cannot afford to be segregated, and although I know that there is much data to show that women who attend single-sex classes ultimately do well, I am not in sympathy with total separation during the critical learning years.

My views no doubt are colored by my own educational background. This included attendance at Girl's Latin School in Boston, a public college-preparatory school restricted to women. Indeed, all public schools in Boston were segregated by gender in those days. As a consequence, I believe that my exposure to mathematics was significantly limited. I felt keenly the lack of a challenge that a coeducational class could have offered in mathematics.

At Radcliffe, where I matriculated following my graduation from high school, we had Harvard professors as our faculty, and they were expected to give us exactly the same courses as they taught at Harvard. My freshman college professor fortunately arranged

for my speeding up my course work and my being allowed the then unheard of privilege of attending Harvard mathematics classes after my sophomore year. I was delighted, but, not surprisingly, I found the going unnecessarily extremely difficult initially until I had adjusted to the far more demanding coeducational environment.

I expect today's young women also struggle with transition to a coeducational culture of graduate school or work once they leave a women's school. Even if the standards of teaching women are comparable, there is always the danger of overly sheltering them in an educational hothouse, though they will have to strive to function in the larger world.

In this respect, again, change has largely eliminated some of the most troubling gender separations. There are, of course, other problems to overcome in single-sex institutions, but in that case, it becomes even more important to avoid depriving female and male students of a chance to interact and to become all the more enriched by observing their differing learning styles. To be an effective learning environment, of course, a coeducational classroom must have a climate friendly to all, with no group given the freedom to dominate or intimidate.

We are making some progress in that direction. For example, an increasing number of presentations at mathematical and general scientific organization meetings are directed toward some of the problems. Also, there are a growing number of current programs that address the issue, and other efforts are evolving as the need becomes recognized.

One particular example, for instance, is that of the Association for Women in Science (AWIS). In 1994, it completed an extensive three-year mentoring project supported by the Arthur P. Sloan Foundation, and AWIS has published a final detailed guide for developing such programs. It is now working on the formulation of a model program offering workable options for institutions committed to enhancing the academic climate for female science faculty. Whereas it deals with the issue in the general area of science, other organizations have introduced programs more specific to mathematics.

The Association for Women in Mathematics (AWM), for one, has developed a substantial cadre of programs that include: workshops for women graduate students and postdoctoral mathematicians at annual meetings of the AMS/MAA and of SIAM; travel grants for women to attend research conferences; and awards recognizing women for their mathematical achievement in various areas. In addition, two bridge programs, one for passage from undergraduate to graduate school and one for passage from first to second year of graduate school, are being prepared for implementation. At the 1995 February meeting of AAAS, the AWM shared its experiences in a session entitled "What works: Successful programs for women in math."

A program that has had an impact since its inception in 1991 is the Mills Summer Mathematics Institute. Designed to increase the number of women who major in mathematics, from all regions of the country, and to encourage the nation's mathematics faculty to direct the most promising ones toward graduate school, the program provides an intensive immersion in mathematics to motivate women and to strengthen their preparation for advanced doctoral study. Its record of having a large proportion of its students doing well in our foremost graduate schools has led to interest, on the part of a number of various other institutions, in introducing summer programs with similar goals.

In collaboration with Mills College, two more programs have been formed,[2] one at St. Olaf and Carleton Colleges and one at the George Washington University. The St. Olaf and Carleton program is primarily aimed at students at the end of their sophomore year. It focuses on conjecture and proof as well as on the use of mathematical software for problem solving. The George Washington University program mainly targets students at the end of their junior year, and concentrates on interactive learning, discovery, and, indeed, when feasible, actual research. Both programs aim to increase the number of women who pursue and obtain advanced degrees in the mathematical sciences and who become active members and leaders of the mathematical community. They emphasize involving active women mathematicians as instructors, teaching assistants, and mentors. These programs have created a network of women mathematicians, graduate students, and undergraduates who support and inspire each other to high achievement.

Ultimately, with a strong, large, and varied group of active mathematicians, there will be greater potential for overall excellence, and the likelihood will be increased for the occasional exceptional mind to emerge, to affect the field dramatically, and thrust it forward. Excellence will then be recognized as a quality that can be achieved without regard to irrelevant factors!

---

[2]This information has been updated to reflect the status of these programs as of July, 1996.

Preparing Undergraduate Women for Graduate School and Beyond[1]

CAROL WOOD

Wesleyan University

With AWM projects, I have found that *who* (is doing the project) determines *what* (project gets done). When I was AWM president, many people of good will would come up and say "you" should do this or that. Sometimes my response would be to point out that the AWM had but one paid employee. The rest of us were volunteers, so whatever got done was done because the members want it and were willing to do it. The AWM now has a bit more support staff, enormously dedicated, but it is still true that initiatives come from the members, and most of the work is still done by volunteers, including the writing of proposals.

I will report here on 1994 activities of the AWM, and then describe an idea for a program that Cora Sadosky and I have been discussing. The AWM was founded in 1971, and overall has about 4,000 members now—over half of them students. It's not good for the budget to have so many student members, but it's great for the organization, especially for the future.

- The AWM newsletter has been published since the AWM's inception. It has for some time been in the hands of the AWM's most capable volunteer, Ann Leggett.

- In 1980 the Noether lectures began. These lectures are given by a senior researcher woman at the annual (winter) AMS-MAA-AWM-NAM meetings.

- In the mid-eighties, AWM Travel Grants were instituted, funded by NSF's mathematics division, DMS. These grants help women to attend research meetings in areas funded by DMS.

- In 1991, with NSF and NSA funding, the AWM began workshops for postdoctoral and graduate students. These workshops take place immediately prior to the winter AMS meetings and the summer SIAM meetings. The workshops bring together women who have recently obtained their PhDs in mathematics and women who are graduate students in mathematics to present their work to a general audience. This is an interesting format, one which encourages the postdoctoral women to talk about their work to non-experts. Graduate students present their work in poster sessions. The workshop also includes non-technical discussions about such topics as career management and funding possibilities. The workshop offers a comfortable setting in which to meet other mathematicians, especially but not exclusively other women, to form valuable contacts, and also to learn about each other's work informally.

- The AWM President organizes a panel discussion at the annual AMS meeting.

- In 1990, the organization instituted the Schafer Prize, a prize for outstanding undergraduate women in mathematics.

- In 1991, the AWM instituted the Hay Award to recognize contributions of women to mathematics education.

- At the AAAS meeting in February 1995, the AWM will hold a panel on "What Counts: Successful Programs for Women in Mathematics."

Now, I turn to my main topic for discussion here, an idea for a bridge program between undergraduate and graduate school. This program would bring together 20 to 25 women, each of whom has been accepted to graduate school, in the summer preceding the start of graduate study.

The program has some very practical aspects plus some mathematical ones. There are three points that have been raised so far in this conference that feed into what I want to say about the purpose of the program. Lenore Blum mentioned that isolation is a big issue. It still is a big issue for a number of young women starting in graduate school in mathematics. Deborah Haimo addressed the comfort level for women studying mathematics—women students really shouldn't have to struggle against anything but hard mathematical problems when in graduate school. Steve Givant mentioned the issue of critical mass.

---

[1]Invited speech given at the Mills Summer Mathematics Institute Conference on Programs for Women in Mathematics, July 14–16, 1994, Berkeley.

The purpose of the program, as Cora Sadosky and I have envisioned it, is to bring together women who have clearly-displayed mathematical talent and mathematical achievement and who have been accepted to graduate programs. We would like to put into place a network for them, before they enter graduate school. The network would serve as a community within which they could talk about issues that they may confront in graduate school. I want to stress that this would be a very upbeat program, not aimed at predicting disasters and bad treatment, but rather to provide signposts about various odd things that the women students may encounter on the route to a PhD degree, as well as strategies for getting the most out of the graduate student years.

The AWM can suggest ways to find out information, and to identify people to whom otherwise awkward questions can be addressed. We would hope to provide a safety net as needed.

It seems to me that there exists at present a tension between whether one should strive to change the mathematical culture, or whether one should learn surviving within the existing culture. There are a number of women mathematicians, especially in my generation, who managed to survive by being totally oblivious, hearing and heeding none of the signals and (un)helpful advice sent our way. Actually, that strategy worked amazingly well in many cases, at least in the short run. But for the long-term prosperity of women mathematicians, it's unlikely to be the best strategy. The present need is to deal with the truly delicate balance between changing and coping.

The workshop would be two to three weeks in length. Women entering graduate school would be eligible to apply. Criteria for selection would include the strength of the applicant's record and qualifications for graduate studies. Some preference would be given to those with limited or no exposure to graduate studies in mathematics, since these students are at special risk during the transition. The program would include some already identified "stars" as well, for whom the experience might provide needed maturation, and certainly would provide cameraderie and raise the mathematical value of the program. The goal is to establish an atypical mathematical culture among the young women, one in which the best of the young women encourage the less well-prepared, rather than intimidate them. Experience with talented undergraduate women indicates that this culture is a comfortable one for many women.

There are other, more profound reasons for wanting to include the very best young women. In *A Room of One's Own*, Virginia Wolfe describes vividly a fictional character, Shakespeare's sister; she had an equal dose of talent, but her life led to tragic obscurity. In the account of the circumstances in which Andrew Wiles solved Fermat's conjecture, one is led to admire the single-minded way in which Wiles attacked this goal. I wish that Wolfe were here now to write about a second fictional woman: Wiles's sister. Any proposal aimed at improving women's place in mathematics must include a place for "Wiles's sister."

The program would consist of workshops and mathematics lectures. Some lectures would give an overview of some exciting research area in mathematics. They would also include sample colloquia and sample graduate course lectures, in which outstanding women mathematicians would talk about their work. One of the purposes of the sample graduate course lectures would be to inform those who had never attended a graduate lecture that one need not understand every word of a lecture in order to take something valuable from it.

The workshops would address a range of cultural issues—graduate versus undergraduate school, time management, how to find out what works for you, fishbowl life, personal safety, freedom from harassment, and the possibilities and pitfalls of community outreach. Graduate women get involved in a lot of things—things that are very worthwhile. These may not always be recognized properly, and may not always work towards long range career success. Women should do what makes sense to them, but there is still room for advice on how to be selective and how to avoid an array of commitments which lead nowhere.

Our main goal in this program would be not only to make the transition easier, but to ensure that the women have resources if things don't go well, and most importantly, to ensure that nothing gets in the way of their being as strong as they possibly can be.

From sitting on various selection committees, I have seen many very strong women mathematicians. I've said time and again: this batch will make it big. But somewhere along the line, something happens. Lives get complicated; personal lives get very complicated—that's another issue and a story for another time. Somewhere between the beginning of graduate school and the tenure decision, a lot of things can and do go wrong for women, things that don't seem to go wrong as often for men. A strong transition from undergraduate to graduate study seems to be of crucial importance in knitting mathematical work into a life that makes sense for women.

I came here to get your reactions, either in this open discussion or later, to learn from you what sorts of things would be useful for a woman, beyond the mathematics she will need to know, to make the transition go smoothly. Our program idea also involves the graduate schools themselves, working with them to find out what they expect, and making sure the communication about expectations is very clear between the young women and the graduate program. Of course if we succeed in this, the effect on the culture will be good for everybody, but we are interested in doing it for women first of all.

Note added in March 1997: this idea as described in 1994 did not go past the preproposal state, for lack of an appropriate funding program. It remains on AWM's list of things to do, awaiting the right mix of personnel and funding sources. Similar ideas have been considered for minority mathematicians. The need persists. Any volunteers?

A View of Mathematics from an Undergraduate Perspective

ANI ADHIKARI AND DEBORAH NOLAN WITH
LUBNA CHUNAWALA, ANGELA HEISING, AND
IRIS WOON

University of California at Berkeley

There are many talented undergraduate women in the field of mathematics, as evidenced by the quality of the more than 1,000 applications to the Mills Summer Mathematics Institute (SMI) from 1991 to 1996. The applicants had a mean grade point average (GPA) of 3.6 (standard deviation SD .3), and the 137 students who participated in the program averaged 3.8 for their GPA (SD .2). Women now make up nearly half the undergraduate math majors in the U.S. (1993 Annual Survey of the American Mathematical Society and the Mathematical Association of America); yet in graduate math programs, only about one-quarter of the American students are women, and at strong well-known mathematics programs the picture is bleaker. For example, in 1994, fewer than 10% of the math PhDs at Berkeley went to women; at MIT, only 1 in 6 of the applicants for math grad school were women (May, 1994); and at the University of Michigan, fewer than 20% of the undergraduate honors math majors were women (Lewis, 1994).

In an effort to answer the questions:

- Why don't women choose to continue their study of mathematics at the graduate level?

- What can the mathematics community do to encourage and prepare undergraduate women to study advanced mathematics?

we surveyed the math majors at the University of California at Berkeley and we studied the applications to the 1993 Mills SMI program. Both groups consist of excellent students, and there are many striking similarities between the quantitative summaries of the Berkeley students and the qualitative statements found in the essays of the Mills SMI applicants. In the analysis presented here, we punctuate the numerical results of the survey with quotes from the SMI applicants on their backgrounds, attitudes toward school and math, and the influences that have shaped their lives.

In our survey we found many similarities between male and female majors. The similarities were in: their high school preparation; GPAs; the reasons for math's being a good major for them; and contact with math professors. However, despite the evidence that more undergraduate women are majoring in mathematics and performing well in their major, important factors were found that can adversely affect these women in their pursuit and successful completion of advanced degrees in the mathematical sciences. The female majors, in comparison to the male majors, showed less self-confidence in their abilities, placed greater importance on the support of teachers and friends, took fewer honors classes, and were less enthusiastic about college math in comparison to high school math.

Two recent studies, Hackett et al (1992) and Chipman et al (1992), support our findings. Hackett et al (1992) surveyed approximately 200 undergraduates enrolled in a school of engineering at a midsized west coast university, and found that, although the women had similar GPA and SAT scores to the men, their perceptions of their academic performance and their career expectations were lower. They conclude that performance accomplishments are very influential in shoring up these unrealistically low expectations, and that it is important for the institution to make efforts to change the learning environment to one that is more positive and supportive. The study by Chipman et al (1992) examined the SAT scores and attitudes toward scientific careers for approximately 1,300 Barnard College women. Their study suggested that the effects of learning math on career outlook is largely mediated by the student's attitude and feelings about math. They call for programs for able college students that build confidence in mathematical ability and increase openness to math-intensive careers.

To increase the representation of women in mathematics we, like Hornig (1979), think the mathematics community can make simple adjustments in the training of mathematicians that would make the discipline more attractive and rewarding for women. Specifically, one way to do this is to design math courses so that students can experience the excitement of discovering mathematical ideas on their own, succeed in tackling difficult problems, and have the opportunity to improve their perceptions of their mathematical abilities. While these changes in the classroom are meant to encourage and interest female students to consider advanced degrees and careers in the mathematical sciences, they benefit male and female students alike.

The Survey

In 1995, under our direction, the students in an upper level statistics course at the University of California at Berkeley (UCB) designed and conducted a survey of the math majors at Berkeley. The main goals of the survey were to find out why these students majored in math, how they are faring at Berkeley, and what their future plans are for graduate studies.

The questionnaire was developed with input from faculty and staff in the Mathematics Department, Statistics Department, and School of Physical Sciences at Berkeley. Some of the questions were modeled after those in earlier surveys on sex-differences among students in the sciences (Donahue 1992, Zappert and Stansburg 1984). The response rate to the questionnaire was very high; 106 of the 115 math majors completed the questionnaire, and many students expressed an interest in seeing the results of the survey. While some students had difficulty remembering their GPA (15% nonresponse rate), virtually everyone was willing to tell us how often they visit their math professor's office, how much time they spent doing math homework in the week prior to the survey, and their plans for graduate work. The response rate for these and the other questions reported here ranges from 94 to 100%.

As there are a little more than 100 declared math majors at UCB, an entire census, rather than a probability sample, was conducted. This precludes the need for determining statistical significance and forming statistical confidence intervals. The summary statistics are for the entire population. In comparison with other, perhaps larger, studies, the results found here are for an entire mini-population. Potential biases from self-selection and nonresponse are therefore not an issue. Of course, the Berkeley math majors need not be representative of the entire population of math majors in the United States. Yet, a detailed study of this group points to basic similarities and differences between male and female math majors that should prove useful in developing successful undergraduate mathematics programs for preparing and encouraging undergraduate women to pursue advanced degrees in the mathematical sciences.

Background

The survey found that the math majors at Berkeley arrive on campus well prepared for collegiate mathematics. Nearly all have taken advanced placement math or calculus, and chemistry or physics in high school (Table 1). After arrival at Berkeley, their achievements remain high, as noted by their average GPA of 3.3 (SD 0.5). These students like math. Four out of five majors would still major in math if given another chance, and over two-thirds are considering graduate work in the mathematical sciences at either the PhD or Master's level.

Due to the system for declaring the major at Berkeley, nearly all of the math majors surveyed are juniors or seniors. Roughly half of these majors are in pure math, the remainder being in applied math. Only 30% of the undergraduate majors at Berkeley are women, with two-thirds of them in applied math.

When asked about their career plans, many mentioned research and teaching (40%). One-third of them specifically mentioned that they planned to be a professor or academic researcher, but most were vague. For nonacademic careers, the computer science field was by far the most popular (15%), followed by finance, law, actuarial science, and consulting, each with about 5%. Many said that they had no plans as of yet (15%), and many did not answer this question (15%).

Why Math?

When asked why mathematics is a good major for them, males and females chose the same reasons with nearly exactly the same frequency (Table 2). They were asked to select up to three reasons in answer to the question:

Why is math a good major for you?

The two top reasons chosen by half of the majors, male and female equally, were that math is a challenging subject matter and that math is fun. These opinions were widely held by the SMI applicants as well. Many SMI applicants reported examples of math problems and projects that excited them and influenced their decision to major in the field. It is not surprising that intelligent and motivated students interested in math, whether female or male, would choose roughly the same reasons for getting

Table 1. Summary Statistics on the UCB Math Majors

	% all majors		% all majors
High School Courses:		Contact with Professor:	
Calculus or AP Math	86	Regularly visit	22
Chemistry	86	Talk about grad school	43
Physics	87		
Graduate plans:			
Possible Masters	63	Double Major	39
Possible Ph.D.	47	Would Major in Math Again	78

Table 2. Why is math a good major for you?

	% of Females	% of Males
Challenging subject matter	53	51
Math is fun for me	47	49
I'm good at math	33	46
Aesthetic appeal	27	30
I know when I have the correct answer	20	18
Lively intellectual community	17	18
Good job opportunities	17	16
I like being able to work alone	13	14

Note these percentages do not add to 100 because multiple responses were allowed.

involved in the field. All of them enjoy math for the intellectual challenges it presents. One Princeton University undergraduate, who applied to the SMI, describes such an experience, which she had in high school:

My junior year in high school I was required to do a science project. I thought it was going to be a drag, but fortunately I was wrong. One day my physics teacher decided to give me a hard time and asked me to integrate $\sqrt{\sin(x)}$ from a to b. Being a little ambitious I said, "Sure, it'll take me a second." After three days and twenty-two pages of trying, I finally said to myself, there's got to be a better way. The only thing I could do was use an approximation method, but it was not satisfactory somehow. In the search for finding the exact answer to the problem, I ended up inventing nine different numerical methods, using every mathematical concept in my heart. I made a science project out of it, and went on to compete at the state science fair.

Doing that science project gave me a small taste of what it's like to do research. I discovered how creative you could be in mathematics, and what a dynamic, artistic subject it is.

Two additional points about Table 2 are of interest. The first is the low frequency for which the majors said that "good job opportunities" was a reason for mathematics's being a good major. For the SMI applicants, we found that, although most report on having loved math from the start, a number of them have tried majoring in subjects other than math. Many were motivated by external circumstances, such as pressure from parents and the seemingly impractical future in math. For example, one applicant enrolled in the College of Engineering at the University of Michigan because

My parents were concerned that I should make a sensible and secure career choice. My father let me know that the prospect for jobs in math is not as good as some other fields, like engineering. I compromised and decided that I could try to apply my enthusiasm and talent for mathematics to something more secure... practicality crept upon my dreams as I grew older.

After two years of mechanical engineering, she switched majors to math. Others report on switching to mathematics from majors such as political science, pre-law, classical studies, physics, and chemical engineering.

It is unfortunate that majors have such a poor impression of the career potentials in mathematics. It may be that they are unfamiliar with the employment opportunities available to them. As described earlier, many of these students are thinking of teaching in the future, or are unsure of their plans. For other majors such as computer science and engineering, the career path is more obvious. Also, this may be why 40% of the math majors at Berkeley are double majors. Computer science is the most popular second major; nearly 40% of the double majors are in computer science, followed by physics, economics, and statistics, which are each between 10 and 15% of the double majors.

The second point is the striking difference between males and females on the selection of the response, "I'm good at mathematics," in answering the question why math is a good major for them (see Table 2). All of the other reasons were chosen with essentially identical frequencies by males and females. But for this choice, there is a 13 percentage-point difference between the sexes. The females are not as willing as the males to acknowledge their abilities in answering this question. One SMI applicant from Humboldt State University in California explains how not internalizing one's abilities can affect the decision to go on to graduate school:

> The woman I now am finds herself holding back even though she knows that she performs well. She finds excuses for her success, thinking that something besides ability must be responsible—the instructor grades too easily, or I just test well.
>
> Until two months ago, I had never considered pursuing a PhD When an instructor suggested it, my first thought was "I could never do that. I couldn't make it." For the first time, I noticed the absence of women in my mathematical experiences. I became aware of the inequalities of which I had heard and never really paid attention. Soon I was angry—not at anyone—but that I should

be disadvantaged because of my sex.

Success in Math

Where male and female majors seemed really to differ was in their responses to the question:

> What has been important to your success in math so far?

Table 3 contains a summary of the responses. These lists are striking in that the females value more highly than the males the individual support from a teacher, friend, or family member. The males attribute their success to reasons that did not involve relationships. They chose talented teachers and hard work far more frequently than the other options.

Again, the students were asked to select up to three reasons in answer to this question. For the males the two main reasons—interesting and talented teachers, and hard work—were chosen by half of them. Alternatively, the females were more diverse in their responses. Their top two reasons were the quality of their high school education and hard work, but a teacher who showed an interest in them was rated essentially as highly. In support of this observation, many SMI students report on role models and mentors who had a strong impact on their mathematical development. One student from the University of California at Davis explains the importance of having an encouraging environment in which to learn mathematics. After having a very difficult time with a teacher in junior high school, she found a teacher in high school who made up for it.

> Luckily in high school I had a teacher who realized my interests and helped to sharpen my abilities. He had confidence in me, which helped me to be secure in what I could accomplish.
>
> My desire to teach began at the same time my interests in math developed. Knowing the importance of a good education, I always admired my teachers. I wanted to be just like them, involved in education.

The value that this student places on the support of a teacher affects her career plans. She wishes to create the same kind of environment that she was

Table 3. What has been important to your success in math so far?

	% of Females	% of Males
Hard Work	47	53
Quality of high school education	47	38
A teacher who showed an interest in me	43	24
Interesting and talented teachers	37	54
Math comes naturally for me	30	36
Family support and encouragement	30	11
Friends' support and encouragement	20	19
Undergraduate research experience	7	7
Job experience	3	3

Note these percentages do not add to 100 because multiple responses were allowed.

fortunate enough to have had. This student was typical of many of those that apply to the SMI.

In addition to a teacher's support, many female students also reported that the support of their family is important to their success in math. This is seen in Table 3, where 30% of the females chose "family support and encouragement" as important to their success in math. In comparison, only 11% of the males chose this as a reason for their success. Additionally, one-third of the female majors at UCB reported having a close relative who works professionally in a mathematics-related field, and only half as many males reported having such a relative. As an example, an SMI applicant from Wellesley College in Stillwater, Minnesota, explains the influence of her mother on her studies and on her career plans.

> Over the years I have observed my mother, my role model and source of encouragement, while being a teacher, continuing to be a student. Even though I did not plan to follow in my mother's footsteps, I have decided I want to be involved in education. Math education seems to nicely combine my interest in working with people with my mathematical interests and experiences. ...I hope to be ...a resource for students as well as maybe being a role model and inspiration for someone in the future.

The benefits of role models and mentors are difficult to measure quantitatively (Merriam 1983, Merriam, Thomas, and Zeph 1987), but we can see from the survey and the SMI applicant statements that mentorship is important to female math majors. We also see that the way in which they learn mathematics is important to them. The process matters.

The process matters to them as students, and it matters to them in their career choices. McMurdy (1992) reports on other studies which have found that female students tend to list service to others as a key consideration in choosing a career, as compared to male students who list salary and job security.

Self-confidence

Widnall (1988) in her AAAS Presidential Lecture, "Voices from the Pipeline", discusses two studies, one at MIT and one at Stanford, of men and women graduate students in the sciences. Both studies found that while the men and women were similar in their preparation and performance in graduate school, they differed greatly in their perceptions of their preparation and the pressure they experience. In the Stanford study (Zappert and Stansbury, 1984), students were asked how often they question their ability to succeed in their field. Twenty-four percent of the females responded that they always or often question their ability, whereas only 9% of the males responded this way.

The math majors at Berkeley were asked this same question, with wording identical to that in the Stanford study (Table 4). Half the female majors reported that they always or often question their ability to succeed in their field, whereas only one-third of the males reported that they do so.

Many of the Mills SMI applicants commented on the change in their confidence in their mathematics abilities that they experienced in the transition from

Table 4. How often do you question your ability to succeed in your field?

	Percent of Females	Percent of Males
Always or often	56	32
Sometimes	27	49
Rarely or never	16	19

high school to college. The women at UCB also expressed less enthusiasm for collegiate mathematics in comparison to high school mathematics. When asked:

> Is math at Berkeley better than math in high school?

60% of the females agreed that it was better. The males were more positive about college math, 80% answered affirmatively. On this point, an SMI applicant from Brown University describes the drain on her self-confidence when she entered university:

> I have spent quite a lot of time in the last few years wondering if I could do mathematics. It did not used to be like that. ...Then I got to college and math class was something entirely different. The problems were no longer calculations. They were proofs. This was more what I understood mathematics to be, but it was hard. If I had felt comfortable enough with my anonymous male classmates to try to work with them, the class would have been easier...I began to doubt if I could master being a real math student, let alone a mathematician.

When a student acts unsure of herself, it can lead her teachers and peers to think poorly of her as well, and the effect snowballs. An SMI applicant from the University of Virginia, who previously participated in a summer research program, aptly describes this problem first hand:

> For two months, I focused on the connection between the ellipse and the Hadamard multiplier norm. While learning about an advanced topic was exciting, I also noticed something about the students, a man and a woman, with whom I was

working. Both of them were good, but their attitudes toward their work... was very different. When responding to questions, the man was quick to answer and confident of his response, right or wrong. The woman, a year his senior in schooling, was less confident of her abilities. Her responses were phrased more like suggestions than answers, and she worried that despite her work, our professor was disappointed in her performance. Their relative levels of self-confidence had an insidious effect on me. While I believed that all three of us were competent, when asked who understood the most math and was making the most progress, I was convinced that it was the male student.

What can the math community do?

The Berkeley math majors and the Mills SMI applicants provide a detailed picture of what it is like to be an undergraduate math major. They conjure up a vast array of students who truly enjoy mathematics, and who like the challenge and the fun in it. To encourage them to pursue advanced degrees in the field, we need to build on this experience. That is, we need to create a classroom environment where students discover mathematical ideas on their own, solve hard problems, and find support in their endeavors. As an SMI applicant from Providence College puts it:

> I have never had a problem with my math courses, but it wasn't until this past semester that I found a professor that allowed me to think in my own way. He *challenged* me in a way I had never been challenged before, and I loved it!

I was so excited I would go home to tell my roommate (who is a history major) about all the things I was learning. ...Math is not some number-crunching refuge for those who cannot relate to other people. Far from it. It involves teamwork and brainstorming—and once you have finally tackled that tough problem, you feel so accomplished—there is nothing like that feeling.

Examples of mathematics seminars and summer programs designed to create an environment like the one this student describes are found in the second and third parts of this manuscript. If the mathematics community pays attention to the classroom climate, then the discipline may be more attractive and rewarding for all its members.

References

Chipman, S., Krantz, D., Silver, R. (1992). Mathematics anxiety and science careers among able college women. *Psychological Science* **3**: 292-295.

Donahue, E. (1992). Preprint. Wellesley College, Amherst, MA.

Hackett, G., Betz, N., Casas, J., Rocha-Singh, I. (1992). Gender, ethnicity, and social cognitive factors predicting the academic achievement of students in engineering. *J. Counseling Psychology* **39**: 527-38.

Hornig, L. (1979). Scientific Sexism. In *Women in Science: Portraits from a World in Transition* (V. Gornick, ed.) 100-109. Ann. New York Acad. Sci., New York.

Lewis, D. (1994). Personal communication.

McMurdy, D. (1992). Gender and the numbers. *Macleins*, Nov. 9, 64-67.

May, P. (1994). Personal communication.

Merriam, S. (1983). Mentors and protégés: A critical review. *Adult Education Quarterly* **54**: 123-144.

Merriam, S., Thomas, T., and Zeph, C. (1987). Mentoring in higher education: What we know now. *Review of Higher Education* **11**: 199-210.

Widnall, S. (1988). AAAS Presidential Lecture: Voices from the Pipeline. *Science* **241**: 1740-45.

Zappert, L., and Stansbury, K. (1984). In the pipeline: A comparative analysis of men and women in graduate programs in science, engineering and medicine at Stanford University. Campus report, Nov 18, 1984, Stanford, California.

Course

Designs

Probability and Stochastic Processes

A. ADHIKARI AND D. NOLAN

University of California at Berkeley

Probability is an excellent topic for a mathematics seminar, because it requires little formal mathematical training, and there are many interesting and hard problems in active research areas that students can start working on quickly.

We have led seminars on probability as part the Mills Summer Mathematics Institute, which is an intensive six-week program for undergraduate women, and at a similar program for minority students held at Berkeley. Although participants in these programs are chosen for their mathematical talent and promise, they come from quite diverse backgrounds (e.g., small colleges, large research institutions, and private universities) and their mathematical preparation varies tremendously. Few have prior knowledge of probability theory. The seminar must thus be designed to accommodate various mathematical backgrounds and to challenge all students. We also structure the seminar to give plenty of practice in written and oral communication of mathematical ideas.

This article contains suggestions on how to organize an introductory seminar in probability and stochastic processes. We provide a series of appendices containing details of the seminar material. They include lists of challenging and interesting problems, a bibliography of research papers suitable for reading by undergraduates, and examples of group projects.

The six-week program is typically organized to get the students quickly ready to work on independent projects. In the first week, the students are given a large number of problems that cover the fundamental properties of probability and expectation and offer preparation for the projects (see the first appendix). A second stage of preparation takes place in week two of the seminar when students work in groups on well-defined mini-projects (see the second and third appendices) that consist of a series of problems that build on each other. For an alternative to the mini-projects, we sometimes have students read an original research paper. The third week of the seminar is spent with the students' presenting the results of their mini-projects or papers, and with the seminar leaders' introducing the areas of the main

research projects. In the last three weeks of the program, students work on their projects, meeting regularly with the instructor and teaching assistant. The class time is spent with students' making progress reports and with special topics.

Essentially, the three basic components of the seminar are: seminar-style teaching; many challenging problems; and individual and group research projects with oral and written presentations. Each is described in turn below.

The Seminar Environment

In the program, students meet with the instructor and teaching assistant three times a week for 90 minutes. In the first few weeks, the beginning of each meeting is used to review previous topics and start new material. New concepts are typically introduced via an example that the class works on as a whole, with the instructor leading the class. Then the students split into groups (pairs or groups of three and four) to work on problems that further develop the material. Typically a time limit is set, and the instructor and TA circulate among the groups, keeping them on track and offering advice. In the last part of the class period, groups choose a spokesperson to present solutions, and assignments are made for the next class meeting.

The students also meet twice a week for an hour with the TA. These meetings are spent working on problems either in the homework or left over from the seminar meeting and on presenting findings orally at the blackboard. Later in the program, this time is used to meet with students to discuss their research projects.

This type of class structure can be used to cover a lot of basic material quickly. Indeed, we have found it crucial to set a rapid pace from the start, and to get students accustomed to the idea of discovering results and proofs for themselves, without the aid of texts. The results are not new, of course, and many can be found in most standard undergraduate texts such as the one by Pitman (1993). But they are new to the students and provide a first taste of research.

The problems in the first appendix were designed to make the students discover fundamental properties of probability and expectation. The students' talent and eagerness led them to demolish these problems in a week, whereas a typical undergraduate class would take almost a month to get this far even with the aid of detailed lectures. Several of the problems provide a workout in real analysis, perhaps the most important mathematical tool used in probability.

Challenging Problems

In addition to problems that reinforce basic concepts, additional challenging problems were handed out on a regular basis. This was done in the last three weeks of the seminar, when students were working on their projects. Some of the problems were related to a group research project. Good sources are Feller (1968), Mosteller (1965), Pitman (1993), and Williams (1991).

The problems may vary in difficulty. Students work those that interest them, and divide up the task of writing solutions. They enjoy the freedom of choosing their assignment. Some work nearly all the problems, while others work in groups on a subset. Occasionally, the class period is dedicated to presenting solutions.

We include here examples of two kinds of challenging problems. The first presents a result in number theory, and the second gives an example of how a classic problem can be generalized and extended in a variety of ways.

The probabilistic method: an example

Number theory is a popular subject among math undergraduates, and the following problems were designed to demonstrate a connection between probability and number theory. The development was modeled after that in Williams's text (1991). Students enjoyed using their new probabilistic tools to establish a well-known result in another branch of mathematics.

The *Riemann zeta function* ζ is defined by

$$\zeta(s) = \sum_{n=1}^{\infty} n^{-s}, \quad s > 1.$$

Consider a positive integer-valued random variable X with distribution given by

$$P(X = n) = \frac{n^{-s}}{\zeta(s)}, \quad n \geq 1.$$

1. Let p be a prime number. Find $P(X$ is divisible by $p)$.

2. First, a definition. Recall that two events A_1 and A_2 are independent if $P(A_1 A_2) = P(A_1)P(A_2)$. Events $A_1, A_2, A_3, \ldots$ are called independent, if for every n and every n-tuple of indices $i_1, i_2, \ldots, i_n$,

$$P(A_{i_1} A_{i_2} \ldots A_{i_n}) = P(A_{i_1})P(A_{i_2}) \ldots P(A_{i_n}).$$

Now back to the problem. Let D_p be the event that X is divisible by p. Show that the events $\{D_p : p$ a prime$\}$ are independent.

3. Here is a general result about the monotonicity of probabilities. Let $A_1, A_2, A_3, \ldots$ be a decreasing sequence of events, that is, $A_1 \supseteq A_2 \supseteq A_3 \supseteq \cdots$. Show that

$$P(A_n) \downarrow P(\bigcap_{i=1}^{\infty} A_i) \text{ as } n \uparrow \infty.$$

4. Use the previous exercises to prove **Euler's formula**: for every $s > 1$,

$$\frac{1}{\zeta(s)} = \prod_{p} (1 - \frac{1}{p^s}),$$

where the product is over all prime numbers.

Variations on a theme: an example

The Buffon needle problem (Kotz and Johnson 1982, vol. 1, p. 324) was solved by the class as a whole. Then variations were given to different groups to tackle and to present to each other.

1. A needle of length l is dropped at random onto a floor marked with parallel lines d apart ($l < d$). What is the chance that the needle crosses the line? (Answer: $2l/\pi d$).

2. What is the expected number of lines crossed if the needle is of length L, where $L > d$? (Answer: View the long needle as k short needles each of length l, where $kl = L$ and $l < d$. Then use indicator functions to find the expectation to be $2L/\pi d$).

3. What is the expected number of lines crossed if the needle is bent? (Answer: The previous solution holds.)

4. A plane curve of length L is inside a circle of diameter d, and a cut is made along a randomly chosen chord as follows: select a distance uniformly from $-d$ to d and an angle uniformly between 0 and π; take the chord to be that which is perpendicular to the diameter corresponding to the chosen angle and which is the chosen distance from the origin. What is the expected number of cuts to the curve? (Answer: An infinitesimal calculus approach similar to that in the previous problems finds: $2L/\pi d$).

Two- and three-dimensional generalizations of these problems from stereology (Baddeley, 1982) can also be solved in groups. For example, the area of a two-dimensional region inside a circle of diameter d can be estimated from the ratio of the length of the region along a randomly chosen chord to the total chord length. In this case, the chord is not generated as above. Instead, an angle is uniformly chosen between 0 and π; a point is chosen at random from the interior of the circle; and the random chord is that which passes through the chosen point along the chosen angle.

Mini-Projects and Research Papers

One of the main goals of a seminar is to introduce the students to research. To lead them into this, instructors can start by assigning substantial problems broken into steps. The aim is to have the students discover the proof of a nontrivial and interesting result that requires a more sustained and deep effort than standard undergraduate work.

The second and third appendices contain examples of two such mini-projects. Each group wrote a small article containing proofs of all their results, and these were photocopied and handed out to all participants. In addition, each group gave a two-hour oral presentation of its project, with every group member being responsible for a section of the material.

The first project is on the arcsine approximation to various distributions connected with the symmetric random walk. The development exploits symmetry and sample path properties, as in the treatment by Feller (1968). Students enjoyed figuring out how to cut and paste bits of sample paths to arrive at the results. Then they had to learn to write out mathematical proofs that their pictorial arguments worked; and lastly, they had to use some real analysis to do the asymptotics.

The second mini-project is on the recurrence and

transience of Markov chains, with special attention to random walks. The project starts with the definition of a Markov chain with a countable state space, and leads students to a proof of the theorem that connects transience and recurrence with the convergence or divergence of a series of n-step transition probabilities. This theorem is applied to study the recurrence and transience of random walks, both symmetric and asymmetric, in one or more dimensions. Like the project on arcsine laws, this one relies heavily on clever uses of elementary probability and real analysis to prove theorems.

An alternative to the mini-projects is to have each student read an original research paper in an area of application of probability. Then each student gives an oral presentation of her reading to the class, writes a detailed abstract for the paper, and contributes a problem related to her paper to the weekly homework assignment. The abstracts can be collected into a scrap book, and copies distributed to all participants. Many students find it exciting to read an original account, rather than prepared textbook material. The papers can cover a wide range of topics. Each student chooses a paper that suits her interests and mathematical preparation. Some students were given supplemental reading for their research papers. An annotated bibliography of research papers that can be read by undergraduates follows the appendices at the end of this article.

Group Projects

Midway through the program, students split into groups to work on projects. The aim is to enable students to discover the field on their own. Group work is structured so that students learn to form genuine collaborations, with each one contributing her strengths, supporting and being supported by the others. They choose the direction in which they want to head, using introductory material covered in the seminar as a starting point.

As they develop their material, they meet regularly with the instructor and teaching assistant, where they receive direction, supplemental reading, and fact sheets on related topics. They make periodic progress reports in class, and each group keeps a notebook of its findings.

For the students to decide which project they want to work on, they are all given an introduction to each topic, and spend a couple of days looking over the materials and making their choice. Included in the fourth appendix are the introductions to five projects. They are very informal, and provide ideas for interesting avenues to explore, but students are urged to investigate other areas and not to limit themselves to those listed here.

Bringing the seminar into the classroom

We have adapted many of these techniques to the regular classroom, and found they can work well for classes that range in size from 20 to 60 students.

Class time remains a combination of lecture and group work. The lecture is used to introduce or wrap up a topic, and in the group work students solve problems in a handout and write their solutions on the blackboard. We rarely ask students to make oral presentations of their solutions. It is usually most effective to have groups working on different problems. For example, to introduce the standard discrete distributions (Bernoulli, binomial, hypergeometric, geometric, negative binomial), students form groups, one for each distribution. Each group works on a concrete problem to find a numeric expression for a probability; e.g., the binomial group finds the chance of three aces in seven rolls of a fair die. When they finish this exercise, they are asked to generalize their result, e.g. to find the chance of exactly k aces in n rolls of a die that lands an ace with chance p. In these exercises, students gain practice formulating a parametric probability model, and are introduced to many of the standard discrete probability distributions through the class presentations.

In a large class, more than one group works on the same problem. Sometimes the class is split into thirds; each third works a different set of problems in small groups. We have developed a series of problems, which are provided in weekly handouts. The handouts include problems that illustrate the main concepts and just-for-fun difficult problems about sleazy gambling joints, winning the lottery twice, and random cuts of spaghetti. By the end of the semester, each student has compiled a set of solutions to approximately 100 problems covering all of the course material. These are in addition to their regular homework problems.

The required material for the course is covered in the first 12-13 weeks of the 15-week semester. The final 2-3 weeks of the course are a combination of review and student presentations of projects. For the projects, they either read an original research article, or work on a set of problems from a subfield of probability that they have not studied. They are given 4-6 weeks to complete their projects. We have found it works best if the students are given a set of questions to guide them in their reading.

Students choose which type of project they wish to work on, and they choose whether to work individually or in a group of up to 5 students. In addition to presenting their project to the class, they also write up their findings in a paper. These projects are very demanding, but we find that the students rise to the challenge and ultimately find a great sense of accomplishment in their work.

Appendix: Introductory Problems to Probability and Expectation

1. The Monty Hall problem. This is a well-known question, and has vexed many people for many years. It arises out of a game show called "Let's Make a Deal," run by Monty Hall. At the end of the show, a contestant stands before three closed doors. Behind one of the doors there is a prize; behind the other two doors there are duds. The contestant doesn't know where the prize is, so she chooses one of the doors at random. No door is opened yet, but it is clear that no matter which door she selects, there is a dud behind at least one of the other two doors. Monty opens one of these other doors to reveal a dud. So now there are two closed doors, one of which is the contestant's originally chosen door. Monty offers her the opportunity to switch her choice of doors to the other unopened door. Should she switch or not? Or does it matter?

2. Symmetry in card shuffling. A deck of cards is well-shuffled, and the cards are dealt one by one. What is the chance that the second card is an ace? What is the chance that the seventeenth card is an ace? What is the chance that the second and third cards are both aces? What is the chance that the seventeenth and fiftieth cards are both aces?

The symmetry in these answers points to a theorem about random permutations. Figure out what the theorem is; then state and prove it.

3. The gambler's rule. Many questions in elementary probability theory were originally posed by gamblers some centuries ago. Here is one of them. A gambler bets repeatedly on an event which occurs with chance $1/N$, where N is a fixed positive integer, usually thought of as large. So for example, he could be betting on the number 0 at roulette; he has chance $1/38$ of winning. Suppose successive bets are independent of each other. How many times should the gambler bet so that the chance that he wins at least once is greater than $1/2$? Gambling experience suggests that the answer is about $2/3$ of N. Is this consistent with your calculation?

4. The birthday problem. There are n people in a class. What is the chance that at least two of them have the same birthday? (What assumptions are you making?) Roughly how big is this chance if $n = 40$? About how many people should there be in the class to make the chance at least 0.5?

5. Another birthday problem. There are n students in your class. What is the chance that at least one of them has the same birthday as yours? Is this problem the same as the birthday problem above?

6. Binomial distribution with parameters n and $p = 1/2$. A fair coin is tossed n times. What is the chance of getting exactly k heads? What is the most likely number of heads? Roughly what is the chance of getting exactly 1,000 heads in 2,000 tosses?

7. Binomial distribution with parameters n and p. Consider n independent trials, each of which results in a success with probability p. What is the chance of getting exactly k successes? What is the most likely number of successes?

8. Negative binomial distribution with parameters k and p. A gambler bets repeatedly on an event which has probability p. The bets are independent of each other, but she decides that she will stop betting as soon as she has won k times. What is the chance that she stops immediately after the nth bet? [Note: the special case $k = 1$ gives rise to the **geometric** distribution with parameter p.]

9. The gambler's ruin problem, with a fair coin. A gambler bets repeatedly on tosses of a fair coin. On each bet, she wins 1 dollar if the coin lands heads, and loses 1 dollar if the coin lands tails. The gambler starts with a dollars in her pocket. She decides to keep betting until either she has won b dollars, or is broke. What is the chance that she ends up broke? [Assume a and b are positive integers.]

10. The gambler's ruin problem, with an unfair coin. Do the gambler's ruin problem again, replacing the fair coin with an unfair coin which lands heads with probability p.

Expectations. Let X be a random variable defined on a countable outcome space Ω, on which there is a probability distribution P. The **expectation** of X is defined as

$$E(X) = \sum_{\omega \in \Omega} X(\omega)P(\omega),$$

provided the series is absolutely convergent; if not, the expectation does not exist.

11. Linearity of Expectation. Let X and Y be random variables defined on Ω, and assume that $E(X)$ and $E(Y)$ exist. For constants c and d, show

$$E(cX + dY) = cE(X) + dE(Y).$$

Though it might seem simple, this property of expectation is of fundamental importance, as you will see in the problems below.

12. Calculating $E(X)$ from the distribution of X. Show that

$$E(X) = \sum xP(X = x)$$

where the sum is over all x in the range of X.

Most texts use this formula to define $E(X)$, by saying something like: "Let X be a random variable with possible values $x_1, x_2, \ldots$. The expectation $E(X)$ is defined as

$$E(X) = \sum_i x_i P(X = x_i)."$$

This is fine, and usually all that's needed in most elementary calculations. But to understand and prove linearity, it is essential to think of the definition on Ω.

13. The method of indicators. Let I_A be the *indicator of the event A*, that is, $I_A = 1$ if A occurs, and $I_A = 0$ if A does not occur. Find $E(I_A)$.

i) A coin which lands heads with probability p is tossed. Use indicators to compute the expected number of heads in n tosses of this coin.

ii) A die is rolled repeatedly. Find the expected number of different faces that appear in the first n rolls.

14. Tail sums. Suppose the possible values of X are $0, 1, 2, 3, \cdots$. Show that

$$E(X) = \sum_{x=0}^{\infty} P(X > x).$$

i) A coin which lands heads with probability p is tossed. Let k be a fixed positive integer. Find the expected number of tosses until the kth head appears. (It's a good idea to think first about the case $k = 1$.)

15. The coupon collector's problem. A cereal company puts one coupon in each of its cereal boxes. There are n distinct coupons. If you collect a complete set of these coupons, you get a grand prize. How many cereal boxes do you expect to buy to get a complete collection? (What assumptions are you making?) Get a good approximation to this expectation for large n.

16. Symmetry in card shuffling, revisited. You deal cards one by one from a well-shuffled deck of cards, until all the aces have been dealt. How many cards do you expect to deal?

Random permutations. Consider the $n!$ permutations of the integers 1 through n. A *random permutation* of 1 through n is obtained by choosing one of these permutations at random so that all $n!$ permutations are equally likely. The next two problems are about such random permutations, though they are couched in traditionally colorful language.

17. Fixed points: the matching problem. There are n letters, labeled 1 through n. And there are n envelopes, also labeled 1 through n. The letters are distributed randomly into the envelopes, one letter per envelope, so that all $n!$ possible arrangements of letters in envelopes are equally likely. Say that a match occurs in envelope i if the letter labeled i falls into the envelope labeled i. Find the expected number of matches.

18. Cycles: the locked boxes problem. There are n boxes. Each box locks itself when slammed shut, and must be opened with its own special key. An annoying person permutes the keys randomly and throws them into the boxes, one key per box. Then she slams all the boxes shut. I want to open the boxes, but am willing to break open only one box. This will yield a key which may open another box, which will yield another key, and so on. Find the expected number of boxes I can open. [Hint: You have solved part of this problem in exercise 7.]

Appendix: Fluctutations in coin tossing—arcsine laws

Let $X_1, X_2, \cdots$ be independent random variables, and for all i, let X_i take the values 1 and -1 with probability $1/2$ each. Let $S_0 = 0$, and for $n \geq 1$, let

$$S_n = X_1 + X_2 + \cdots + X_n.$$

The sequence S_n is called a simple symmetric **random walk** starting at 0.

We will think of $(S_0, S_1, S_2, \cdots, S_n)$ as a polygonal path, whose segments are

$$(k - 1, S_{k-1}) \to (k, S_k).$$

Preliminaries.

1. How many possible paths are there from $(0, 0)$ to (n, x)?

2. Find $P(S_{2n} = 0)$.

3. Stirling's formula—almost! Let n be a positive integer. Show that there is a constant c such that

$$n! \sim cn^n e^{-n} \sqrt{n}.$$

[Hint: compare $\log(n!)$ with $\int_1^n \log x \, dx$. It helps to draw a picture.]

In fact $c = \sqrt{2\pi}$. It takes a little work to prove this, but you can assume it.

4. Let $u_{2n} = P(S_{2n} = 0)$. Show that

$$u_{2n} \sim \frac{1}{\sqrt{\pi n}}.$$

What does this say about the statement, "In the long run you expect about half heads and half tails"?

Arguments by translation and reflection.

5. Warm up: the reflection principle. Let z and w be positive integers. Show that the number of paths from $(0, z)$ to (n, w) that touch or cross the x-axis is equal to the number of paths from $(0, -z)$ to (n, w).

6. Show that

$$u_{2n} = P(S_1 \geq 0, S_2 \geq 0, \cdots, S_{2n} \geq 0).$$

To do this, consider all paths of length $2n$. Let the paths that end at level 0 be called "Type A" paths, and the ones that never get below the x-axis be "Type B." The idea is to establish a one-to-one correspondence between paths of these two types. Here is how to start. Take a Type A path. If it never gets below the x-axis, leave it alone. If it does, let $(k, -m)$ be the coordinates of the leftmost lowest point of the path. Make this point your new origin. Take the initial portion from $(0, 0)$ to $(k, -m)$, reflect it about the vertical line $x = k$, chop it off, and add it on to the end of the path. Draw lots of pictures!

7. Show that

$$2P(S_1 > 0, S_2 > 0, \cdots, S_{2n} > 0)$$
$$= P(S_1 \geq 0, S_2 \geq 0, \cdots, S_{2n} \geq 0).$$

8. Show that

$$P(S_1 \neq 0, S_2 \neq 0, \cdots, S_{2n} \neq 0) = u_{2n}.$$

Arcsine laws.

9. Let $k < n$. In terms of the u's, find $\alpha_{2k,2n}$, defined by

$$\alpha_{2k,2n} = P(S_{2k} = 0, S_{2k+1} \neq 0,$$
$$S_{2k+2} \neq 0, \cdots, S_{2n} \neq 0).$$

10. Get a good approximation to $\alpha_{2k,2n}$ and sketch a graph of this approximation as a function of k for fixed n. What does the graph say about the time of the last zero in a random walk of $2n$ steps?

11. An arcsine law. Let x be between 0 and 1, and let the random walk go for a very long time. Show that the chance that the last zero occurs before a fraction x of the total time is approximately

$$\frac{2}{\pi} \arcsin \sqrt{x}.$$

12. Arcsine law for the time of the first maximum. Consider a random walk of n steps. Say that the first maximum occurs at time k if

$$S_0 < S_k, S_1 < S_k, \ldots, S_{k-1} < S_k,$$

and

$$S_{k+1} \leq S_k, S_{k+2} \leq S_k, \ldots, S_n \leq S_k.$$

Show that the time of the first maximum also follows an arcsine law. [Hint: Consider the "dual" random walk; the first step is X_n, second step is X_{n-1}, etc.]

Appendix: Recurrence and Transience in Markov Chains

A stochastic process $\{X_0, X_1, X_2, X_3, \ldots\}$ is called a *Markov chain* if for each n, the conditional distribution of X_{n+1} given $(X_0, X_1, \ldots, X_n)$ is the same as the conditional distribution of X_{n+1} given just X_n.

Assume each X_i takes values in a countable set S, called the state space of the process. For ease of notation, we will assume that S is the set of integers. Then $\{X_0, X_1, X_2, \ldots\}$ is a Markov Chain if for every n and all integers $i_0, i_1, \ldots, i_n, i_{n+1}$,

$$P(X_{n+1} = i_{n+1} | X_0 = i_0, \ldots, X_n = i_n)$$
$$= P(X_{n+1} = i_{n+1} | X_n = i_n).$$

The probability on the right hand side is called a *transition probability*, because it is the probability that the process makes a transition to state i_{n+1} at time $n + 1$ given that it was in state i_n at time n.

In general, the transition probability

$$P(X_{n+1} = j | X_n = i)$$

depends on i, j, and n. But in many interesting examples, it depends only on i and j. Such transition probabilities are called *stationary*, and you can write the probability as a function of just i and j:

$$P(X_{n+1} = j | X_n = i) = p_{ij}.$$

As the notation suggests, these transition probabilities can be arranged in a *transition matrix* $\mathbf{P}$, whose (i, j)th element is p_{ij}.

In what follows, $\{X_0, X_1, X_2, \ldots\}$ is an integer-valued Markov chain with stationary transition matrix $\mathbf{P}$. The process is said to start at time 0 with value X_0.

1. The n-step transition probabilities. Let $p_{ij}^{(n)}$ be the probability that the chain is at state j at time n, given that it started at state i. Show that $p_{ij}^{(n)}$ is the (i, j)th element of the matrix $\mathbf{P}^n$. [To avoid irritation later, define $\mathbf{P}^0$ to be the identity matrix, and check that this definition makes intuitive sense.]

2. First-hitting probabilities. Let $f_{ij}^{(n)}$ be the probability that the chain visits state j *for the first time* at time n, given that it started at state i. That is, $f_{ij}^{(n)}$ is

$$P(X_1 \neq j, X_2 \neq j, \cdots, X_{n-1} \neq j, X_n = j | X_0 = i).$$

a) Which is bigger: $f_{ij}^{(n)}$ or $p_{ij}^{(n)}$?

b) Let f_{ij} be the probability that the chain ever visits state j, given that it started at i. Find a formula for f_{ij} in terms of the $f_{ij}^{(n)}$'s.

Recurrence and Transience. A state i is called *recurrent* if $f_{ii} = 1$, and *transient* if $f_{ii} < 1$.

So i is recurrent if, given that the chain started at i, it is certain to return to i. If there is some chance that it does not return to i, then i is transient.

Consider the simple random walk determined by tosses of a coin which lands heads with probability p. Suppose the walk starts at 0. For which values of p do you think the state 0 will be transient? For which values will it be recurrent? [For now just say what your intuition tells you—by the end of this project you will have a rigorous answer!]

Back to the Markov chain. For brevity, write $P_i(A)$ for the probability of A given that the chain started in state i. That is, $P_i(A) = P(A|X_0 = i)$.

The main theorem of this project set says:

Transience of the state i is equivalent to

$$P_i(\text{infinitely many visits to } i) = 0,$$

and is also equivalent to

$$\sum_n p_{ii}^{(n)} < \infty.$$

Recurrence of the state i is equivalent to

$$P_i(\text{infinitely many visits to } i) = 1,$$

and is also equivalent to

$$\sum_n p_{ii}^{(n)} = \infty.$$

Before you prove this theorem, use it to check your intuitive answer to the question about random walks above.

Exercises 3–6 constitute a proof of the theorem.

3. In terms of the f's defined in Exercise 2, find a formula for the probability that the chain makes at least m visits to j, given that it started at i. Hence find a very simple formula for P_i (infinitely many visits to j), in terms of the f's. Specialize this to the case $j = i$.

4. Show that $\sum_n p_{ii}^{(n)}$ is the expectation of a certain random variable. Deduce that

$$\sum_n p_{ii}^{(n)} < \infty$$

implies

$$P_i(\text{infinitely many visits to } i) = 0.$$

5. Show that $f_{ii} < 1$ implies $\sum_n p_{ii}^{(n)} < \infty$, in the following steps.

a) Show

$$p_{ij}^{(n)} = \sum_{s=0}^{n-1} f_{ij}^{(n-s)} p_{jj}^{(s)}.$$

b) Specialize to the case $j = i$, and sum both sides above, to show

$$\sum_{t=1}^{n} p_{ii}^{(t)} \leq \sum_{s=0}^{n} p_{ii}^{(s)} f_{ii}.$$

c) Now complete the exercise.

6. Show that the theorem has been proved!

7. More than one dimension. Consider a simple symmetric random walk in d dimensions, defined as follows. Start at the origin. At each stage, toss d fair coins, and change the ith coordinate by $+1$ or -1 according to whether the ith coin lands heads or tails. Show that the origin is recurrent in 1 and 2 dimensions, but transient in 3 or more dimensions.

8. Recurrent (or transient) chains. Call a Markov chain *irreducible* if every state can be reached with positive probability from every other state—that is, if for all i and j there is some n for which $p_{ij}^{(n)} > 0$.

a) Give an example of a Markov chain which is not irreducible.

b) Show that in an irreducible chain, either all states are recurrent, or none is. Thus, for example, we can say, "the random walk is recurrent" once we have checked that the origin is a recurrent state.

Appendix: Group Research Projects

Martingales

Loosely speaking, a martingale is a process in which the expected value of the process tomorrow, given all information up through today, equals the value of the process today. In that sense, it can be thought of as your sequence of fortunes in a fair game, though as you have seen, this doesn't necessarily imply that you'll break even in the long run!

Martingales have beautiful properties: under mild regularity conditions they converge in various ways, and there are bounds on how wild they can get. Many interesting problems, apparently unrelated to martingales, can be solved using Optional Stopping Theorems (OST), which say the following: it's easy to check that the expected value of a martingale at every fixed time is the same, but the same is also true at certain well-behaved random times, known as stopping times. This allows you to compute all kinds of probabilities and expectations, as you will see.

It is often fun to "hunt the martingale" in problems so that the powerful martingale methods can be used in solutions. See for example the connection with the branching processes project.

The study of martingales is central to modern probability theory. To do it right, you have to be careful about exactly what is meant by "expected values given all the information up through today," etc. This involves analysis and measure theory, which are good areas for you to get acquainted with anyway.

Suggested starting place. Read the lecture note sketches, and for the moment don't worry that there are gaping holes in them (you'll plug those holes later). Trust that the OST is true and use it to get the expected duration of the game in the gambler's ruin problem, and the expected amount of time a monkey will take to type ABRACADABRA. This last is not as silly a problem as it might seem; similar issues become very important in data compression. Once you have done this, read Williams (1991) for a proof of the OST.

References. There are any number of books about martingales. A good first book is Williams (1991). Eventually, you should be able to read Li (1980) on martingales in pattern matching.

Generating Functions and Branching Processes

This concerns a particular population growth model: start with one individual, who then has a random number of offspring according to a certain distribution; each offspring is a clone of the parent, in the sense that each offspring in turn has a random number of offspring, with the same distribution as the offspring distribution of the original parent. All individuals reproduce independently of each other. It is natural to draw the "family tree," hence the name "branching process."

Questions of interest concern the long-run behavior of the population: does it die out? does it explode? And so on. Answers are very elegant.

But it helps to have some math technique, namely the theory of generating functions. This is a general technique involving power series, which often makes short work of otherwise intractable problems. For example, you can tackle the unfair-coin random walk, which doesn't have the pretty symmetries of the fair-coin walk. The methods are analytical, rather than probabilistic, but well worth learning as they pop up in all kinds of places.

Suggested starting place. Glance through the branching process lecture, first ignoring the math, and try to sift out the results. You might try taking a very simple offspring distribution and see if you can figure out yourself whether your population will die out. Then read Chapter 0 in Williams (1991). Then start learning about generating functions.

References. Chapter 0 in Williams (1991), referred to above, is a good introduction to the branching process, and it provides motivation for the future use of measure theory. Williams also cites many excellent works on branching processes. Wilf (1994) is good too for learning about generating functions.

Connections with other projects. There's a useful martingale hidden in the branching process.

Limit distributions of Markov chains

Suppose you run a Markov chain for a long time. How is it going to behave? You can guess at some of the behavior: because of the Markov property which says the "previous step is the only one that matters," the chain should somehow forget where it started. In fact it does, under certain conditions, and it has a nice limit distribution, which is easy to compute and has useful properties.

The big theorem is that under certain conditions on the one-step transition probabilities p_{ij}, the n-step transition probabilities converge: $\lim_n p_{ij}^{(n)} \to \pi_j$ as $n \to \infty$. Notice that the limit does not depend on the starting state i. It is then easy to check that the vector π satisfies the equations $\pi\mathbf{P} = \pi$, and hence you have a way of computing it. Explicitly computing π in special cases is a lot of fun and yields interesting results.

The "invariant distribution" π has many other interpretations and uses. It allows you to compute things like the expected number of visits to i before the next visit to j, the long-run fraction of time the chain spends at any given state, etc.

There is more than one way of proving the big theorem; after all, in some sense it's just a theorem about matrices, and shouldn't involve any probability. But the proof outlined in the project is very "probabilistic," and uses a method called "coupling." This is a general method which works as follows. Suppose you are trying to prove that your Markov chain has a certain asymptotic behavior. And suppose you know of some other "nice" Markov chain that clearly has that same asymptotic behavior. Set them both running, and try to show that the two paths are bound to meet somewhere. If you can do this, you are home free, because after the meeting-point your Markov chain is probabilistically equivalent to the other one, and therefore must have the same asymptotic behavior!

Suggested starting place. The statement of the big theorem is at the bottom of page 10 of Williams. You'll quickly see that you need to learn some terminology, etc. At this point you can simply start reading from page 1, and mark the places where things aren't proved. Most are proved in the appendix, but it's best to start filling in the gaps on your own. It will be good preparation for proving the big theorem.

References. Kendall (1994) is a nice book. Asmussen (1987) on applied probability has a clear exposition of the coupling proof. Karlin-Taylor (1975) is a standard reference.

Card shuffling and random permutations

How many times do you have to shuffle a deck of cards until it's random? Yeah, right, seven. But what does the question mean, and why seven? This is addressed in the two papers, Aldous and Diaconis (1986) and Bayer and Diaconis (1992), and followed up in a couple of others.

In Aldous and Diaconis (1986), you start with a deck of n cards, arranged in order from 1 to n. Now let's use a very basic shuffle: pick up the top card and insert it at random into the deck. Repeat with the new top card, and so on. Every time you shuffle, the configuration of the deck follows a probability distribution on the space of $n!$ permutations. Clearly, when the number k of shuffles is small, this distribution is far from uniform. As k gets large, you hope that the distribution converges to uniform, in some sense.

In what sense? First you have to define a distance between the distribution of the state of the deck after k shuffles, and the uniform distribution on $n!$ permutations. This is called the total variation distance $d(k)$, and is quite a simple object in our problem. The miracle about the function d is that it is pretty large up to a certain k, then drops sharply to very close to zero. That is why you can reasonably say, "After so many shuffles, the deck is pretty close to random." No, the cut-off k is not 7 for this shuffle on 52 cards; it's more like 250! But it's 7 for the riffle shuffle which is far less dorky than "top in at random." That's what Bayer and Diaconis (1992) is about.

Suggested starting place. The Introduction to Aldous and Diaconis (1986). Note the intimate connection with a problem you did in the first week!

References. You'll get a fair way without any additional references.

Connections to other projects. Note the ubiquitous stopping times. The project on martingales is related to this project as is the Markov chains project. The Markov chains project involves showing that as the number of shuffles tends to ∞, the distribution of the state of the deck tends to uniform, but the heart of this project is the rate of convergence.

Random Permutations, Cycles and Hungarian Maps

Random permutations were introduced in the letter matching problem and the locked box problem. Suppose there were 7 locked boxes and keys arranged as follows

$$\begin{pmatrix} \text{Box } & 1\ 2\ 3\ 4\ 5\ 6\ 7 \\ \text{Key } & 5\ 1\ 7\ 4\ 6\ 2\ 3 \end{pmatrix}.$$

A cycle representation for this permutation is:

$$(1562)(37)(4).$$

There are 3 cycles, one of length 4, one of length 2, and one of length 1. The cycle of length 1 represents a match of box and key. Another way to express the cycle representation for the permutation that does not require the use of parentheses is to use the convention that each cycle is written so as to end in its smallest number, and cycles are written in the order of increasing smallest last numbers. This representation is called the Hungarian map, and for the above permutation the Hungarian map is 5621734.

It may be interesting to investigate the properties of the number of cycles of length 1, the length of the cycle that contains 1, or the total number of cycles in a random permutation.

What is the chance the permutation is an involution, i.e., that it contains only cycles of length 1 and 2? What is the chance that all cycle lengths are divisible by d? To answer these questions, it may be helpful to consider the cycle vector for a permutation. It is the vector $(l_1, l_2, \ldots, l_n)$, where l_i is the number of cycles of length i in the permutation. For our example, the cycle vector is $(1,1,0,1,0,0,0)$. Notice that $\sum i l_i = n$. Can you show that the probability mass function for the cycle vector is:

$$\mathbf{P}(l_1, \ldots, l_n) = \Pi_{i=1}^{n} \frac{1}{(l_i)!\ i^{l_i}}?$$

Can you find the probability generating function for the cycle vector? To do this you will need to learn about multivariate probability mass functions and generating functions.

Another interesting avenue to explore is the connection between cycles and records. Consider the independent records, R_1, R_2, ..., R_n, where R_i has a discrete uniform distribution on $\{1, 2, \ldots, i\}$. Try generating the Hungarian map in reverse order by first choosing the R_n^{th} element from the ordered set $\{1, 2, \ldots, n\}$; then choose the R_{n-1}^{th} element from the $\{1, 2, \ldots, n\} - R_n^{th}$ *element*, and so on. What do you find?

There is also a connection between Polya's urn and cycles. Consider the urn with 1 red and 1 black ball. Mark the red ball with the number 1. Then draw a ball at random from the urn, and return it along with an additional ball of the same color to the urn. But, before putting the new ball into the urn, mark it with the number 2. Continue drawing balls in this fashion. That is, on the ith draw, a ball is chosen at random from the urn, then the ball is returned to the urn along with another ball that is the same color and that is marked $i + 1$. What is the connection between the numbers on the red balls in the urn and cycles? What happens to the proportion of red balls in the urn as the number of draws goes to infinity?

Suggested starting place. The problems in the first week's handout (see Appendix) are a good place to start. Can you rephrase these questions in terms of cycles and Hungarian maps?

References. There are very few references on this topic. You may want to look at some problems in Billingsley (1986) and Feller (1968). This project is modeled after a seminar by Diaconis and Pitman held in 1987, and draws upon their unpublished lecture notes.

Connections with other projects. The card shuffling project is about finding out how many shuffles it takes to reach a random permutation. Polya's urn has a hidden martingale in it; and there is an interesting connection to Markov chains.

Annotated Bibliography

Provided here is an annotated list of research papers that are suitable for undergraduates to read.

Aldous, D., and Diaconis, P. (1986). Shuffling Cards and Stopping Times. *American Math Monthly* **93**: 333–348. The first two sections of this paper examine top-in-at-random shuffling. Bounds are found on the number of shuffles needed for the cards to be in random order. To solve the problem, students need to work with random permutations, variation distance, stopping times, and Chebychev's inequality.

Diaconis, P., and Freedman, D. (1977). How to round percentages. *J. Amer. Statist. Assoc.* **74**:359–364. This paper looks at probability models for generating tables of percentages that add to

100%, exactly. Students work with the multinomial distribution, rounding transformations, and laws of large numbers. Theorem 1 can be easily followed, and the later sections have interesting small sample results that can be programmed on a computer.

Dorfman, R. (1943). The detection of defective members of large populations. *Annals of Mathematical Statistics* **14**: 436–40. This self-contained paper is an application of two-stage testing for a disease by pooling samples of blood. It requires work with conditional expectations and discrete probability. For further reading, a three-stage procedure is discussed in Finucan (1964). The blood testing problem. *Applied Statistics* **13**: 43–50.

Ferguson, T.S. (1965). A characterization of the geometric distribution. *Amer. Math Monthly* **72**: 256–60. This technical article develops a characterization of two independent discrete random variables that can only hold if they have geometric distributions. It works with the independence properties of the minimum and the difference of two random variables.

Halley, E. (1693). *Philosophical transactions—giving some account of the present undertakings, studies and labours of the ingenious, in many considerable parts of the world.* Halley's work on the uses of life tables is reprinted in *World of Mathematics* (1956). An excellent supplement to this reading is pages 131–143 in: Hald, A. (1989). *A History of Probability and Statistics and Their Applications Before 1750.* Wiley, New York, 131–143. There are many interesting problems to work on from Halley's methods, such as the relation between tail sums and expectation.

Hendricks, W. (1972). A single-shelf library of N books. *J. Applied Probab.* **9**: 231–233. Some additional background reading in Markov chains is required. A simple move-to-the-front scheme for reshelving books is put in a Markov Chain framework. The stationary distribution is determined by induction.

Kingman, J. (1982). The thrown string (with discussion). *J. Royal Statist. Soc., Ser. B* **44**: 109–138. A rigorous treatment of the extension of Buffon's needle discussed earlier.

Knuth, D. (1984). The toilet paper problem. *Amer. Math. Monthly* **91**: 465–470. This problem is related to the classical ballot problem (Feller, 1968). Recurrence relations are used to find a generating function, which leads to large sample behavior for the problem.

Mendel, G. (1865). *An Experiment in Plant Hybridization.* An account of Mendel's work is found in *World of Mathematics* (1956). Mendel's paper is quite elementary, but it offers many avenues for exploration. Feller (1968, sec. V.8) contains a variety of problems including some related to the Hardy-Weinberg equilibrium and sex-linked traits. The question of how well Mendel's data fit his model is discussed in Chapter 25 of Freedman, D., Pisani, R., Purves, R., and Adhikari, A. (1991). *Statistics.* Norton, New York.

Moran, P.A.P. (1951). A mathematical theory of animal trapping. *Biometrika* **38**: 307–11. This paper develops a probability model for estimating the size of an animal population using maximum likelihood estimation and Stirling's approximation.

Newell, G.F. (1959). A theory of platoon formation in tunnel traffic. *Operations Research* **7**: 589–98. This paper constructs a probability model for the flow of traffic through a tunnel; it includes finding the distribution of a maximum of independent random variables.

Ramakrishnan, S., and Sudderth, W.D. (1988). A sequence of coin tosses for which the strong law fails. *American Mathematical Monthly* **95**: 939–41. Some measure theory, such as that in Chapter 1 of Billingsley (1968), is required for this paper.

Student (1906). On the error of counting with a haemacytometer. *Biometrika* **5**: 351–60. This paper derives the Poisson process in the plane from first principles. It offers good practice with infinite series and moment generating functions. Any supplemental reading from an undergraduate text on the Poisson process is helpful for synthesizing the results presented here.

Tversky, A., and Gilovich, T. (1989). The cold facts about the hot hand in Basketball. *Chance* **2**, issue 1, 16–21. Plus additional articles in *Chance* by Larkey, Smith, and Kadane, issue 4, 22–30, and Tversky and Gilovich, issue 4, 31–34. This series of articles provides a very accessible analysis of streak shooting from actual basketball games and designed experiments. Additional reading on the distribution of run tests helps focus the project. Lehmann, E. (1975). *Nonparametrics: Statistical Methods Based on Ranks.* Holden-Day, San Francisco.

Vajdas, S. (1956). Theory of games in *The World of Mathematics* **2**: 1285–1293. Although Vajdas's account of zero-sum games is not an original research paper, it provides a clear concise introduction to the area, and contains interesting examples to work out.

Warner, S.L. (1965). Randomized response: a survey technique for eliminating evasive answer bias. *Journal of the American Statistical Association* **60**: 63–69. This paper contains a good example of conditional probability. It also uses the method of maximum likelihood to estimate the parameters in the binomial distribution.

References

Aldous, D., and Diaconis, P. (1986). Shuffling cards and stopping times. *American Math Monthly* **93**: 333–348.

Asmussen, S. (1987). *Applied Probability and Queues.* Wiley, New York.

Baddeley, A. (1982). Stochastic geometry: An introduction and reading-list. *International Statistics Review* **50**: 179–193.

Bayer, D., and Diaconis, P. (1992). Trailing the dovetail shuffle to its lair. *Ann. Appl. Probab.* **2**: 294–313.

Billingsley, P. (1986). *Probability and Measure* (2nd ed.). Wiley, New York.

Feller, W. (1968). *An Introduction to Probability Theory and Its Applications*, Vol. I. Wiley, New York.

Karlin, S., and Taylor, H. (1975). *A First Course in Stochastic Processes.* Academic Press, New York.

Kendall, M. (1994). *The Advanced Theory of Statistics*, 6th edition. Academic Press, New York.

Kotz, S., and Johnson, N.L. (1982). *Encyclopedia of Statistical Sciences.* Wiley, New York.

Li, S.Y. (1980). A martingale approach to the study of occurrence of sequence patterns in repeated experiments. *Ann. Probab.* **8**: 1171–1176.

Mosteller, F. (1965). *Fifty Challenging Problems with Solutions.* Dover, Mineola, N.Y.

Newman, J.R. (1956). *World of Mathematics.* Simon and Schuster, New York.

Pitman, J. (1993). *Probability.* Springer-Verlag, New York.

Wilf, H. (1994). *Generating Functionology.* Academic Press, New York.

Williams, D. (1991). *Probability with Martingales.* Cambridge University Press, New York.

Hyperplane Arrangements

Hélène Barcelo

Arizona State University, Tempe

Introduction

What follows is an outline of a six-week seminar on hyperplane arrangements. The purpose of this seminar was to introduce talented undergraduate students to a very active area of research. The participants had diverse mathematical backgrounds even though all of them had completed the first two years of college. The topic of hyperplane arrangements turns out to be quite suitable for such a seminar. Indeed, it does not require a lot of mathematical background, and yet in six weeks the students are given a real opportunity to work on hard problems and to get an honest experience in research.

Three main ideas guided me throughout the seminar. Firstly, the students should discover and prove new results, and should start working on their own as soon as possible. Secondly, I encourage teamwork, with students working in groups of two to four. For this, I wanted the different topics of research to be sufficiently interrelated so that every group would have a fair chance of fully grasping the results of the other groups. Lastly, there should be a common topic through which I would introduce the concepts they needed for their particular projects.

Coxeter arrangements are ideally suited for these goals. Since there are many Coxeter groups, each team can pick a different group, while the symmetric group is a perfect common topic of study. A hyperplane arrangement is a finite collection of co-dimension 1 subspaces of a vector space. We dealt (mostly) with real vector spaces. Certain hyperplane arrangements may be studied in terms of the group generated by reflections across these hyperplanes. We concentrated only on those arrangements where this associated group is a finite Coxeter group. Conversely, to each finite Coxeter group, represented as a group of reflections, one may associate the arrangement of hyperplanes fixed by these reflections. The braid arrangement, which corresponds to the Coxeter group A_n, has been extensively studied, and is a good prototype for the whole subject. The properties of this arrangement were developed in the lectures. The arrangements corresponding to the Cox-

eter groups of type B_n and D_n are eminently suitable for small groups of students. Unfortunately, there are no other classical Coxeter groups, so an exceptional group had to be chosen as a third topic. So, F_4 was selected, for it is sufficiently small to be tractable but not too simple to be uninteresting, as the dihedral groups would have been. At first, I was somewhat worried that the group choosing to work on F_4 would not be able to prove as much as the other students could. It turned out quite differently: the group working on F_4 gave a very nice proof not found in the literature.

In the next section, I describe the weekly contents of the seminar. I met with the whole class for 90 minutes three times a week. The TA, Jessica Standon, met with them the remaining two days of the week. Except for the first hour of the first meeting, I lectured only when the students asked me to do so; instead, I provided them with lecture notes. These notes, highly inspired by Orlik and Terao (1992) (and with some small passages taken directly from this book), were intended to present some of the theory of hyperplane arrangements in as simple and brief a manner as possible. I wanted the students to take an active part in learning. Thus I left many blanks in examples and proofs to be filled by the students to check how well they understood the material. They also had many exercises to do. During their Tuesday and Thursday meetings with the TA, they usually worked on these notes and exercises. It should be noted that the following description is only a guide; it does not contain sufficient detail to be at all comprehensive. The notes and exercises for each week of the seminar follow the overview in the second section.

Overview

First Week

The goal of the first meeting was to give them an overview of the mathematics involved in hyperplane arrangements. For this I read with them some notes, which I reproduce in the next section. Much of this was too advanced for them at the time; nevertheless, it constituted a basis to which they could refer later during the seminar. As I was reading this text with them, I was focusing on the intuitive meanings of the different concepts. These notes are taken directly from the introduction of Orlik and Terao's book (1992) and are a summary of the topics we were to cover during the seminar.

After the first hour of introduction, the students were given a set of problems (see next section) to work and report on during the next meeting. This series of exercises serves several purposes. One is to have them think about arrangements in $\mathbb{R}^2$ and $\mathbb{R}^3$ (no. 1, 2). Next we look at arrangements of lines that are not in general position. They are led (no. 3–8) to discover Roberts's heuristic arguments (Roberts, 1889) and realize why these arguments do not constitute a correct proof (no. 9). They were then given a paper by Wetzel (1978) which contains three different proofs of Roberts's formula. They were organized in three groups, and each group had to present a proof to the rest of the class in the next Friday meeting (no. 10).

During the Wednesday meeting, the students were given the opportunity to discuss with Jessica and me their solutions to the problems and the proofs they were to present on Friday.

On Friday, the students (divided in three groups) presented their proofs of Roberts's formula. Each member of the groups went to the blackboard, which was their own decision, and they seemed thrilled by this experience. This is a valuable experience for them, and it also allows the instructor to identify the strong and the weak points of each student. On the other hand, it does take up a lot of class time; I was unable to keep them to their allotted time, and the three presentations together took almost the entire class time.

Second Week

Monday and Wednesday were spent learning quite a lot of material: a formal definition of hyperplane arrangements, and their defining polynomials, posets, rank function, join and meet operations, lattices, modularity, Möbius function, and characteristic polynomial. They were also introduced to the arrangements they would study, namely A_{l-1}, B_l, D_l, and F_4. They learned in the end that to find the number of regions (chambers) formed by an arrangement, one needs to calculate the value of its characteristic polynomial at -1. The proof of this theorem was to be presented in class on Friday by one of the groups. There were two other projects to be presented on Friday. One group presented the notion of the matroid that can be associated to an arrangement (Ziegler, 1992). They also discussed some of the pros and cons of matroids versus intersection lattices. The third group presented Za-

slavsky's (1975) notion of relatively bounded faces of an arrangement in order to be able to count the number of bounded regions. Lastly, there was one student who preferred to work alone. She showed (see for example Orlik and Terao, 1992) how the Möbius function of a poset is a generalization of the number theoretic Möbius function.

As I mentioned in the introduction, I did not lecture unless asked by the students. The notes given to them in the second week are in a later section. The words in italics represent concepts they should know, and they were responsible for writing the formal definitions of these in their notebooks. The boldface words represent the new definitions.

Third Week

The students were still struggling with the notions of rank function as well as that of meet and join, and on Monday morning they asked me to lecture on these topics. I did, and it took us the whole period to clarify their misunderstandings. On Wednesday, they worked on the set of notes given below, which are concerned with the Poincaré polynomial of an arrangement as well as the notion of modularity.

On Friday, I showed them the bijection between the lattice of the braid arrangement and the partition lattice. From then on, when dealing with the braid arrangement we always used the partition lattice Π_n. On that day, one group also presented the proof of the deletion-restriction theorem which yields a recursive algorithm for computing the Poincaré polynomial of a lattice. As I mentioned earlier there were three groups of students working on the B_n, D_n, and F_4 arrangements, respectively. During this week, students started to work on arrangements for their group; they looked for modular elements, and the groups working on B_n and D_n were able to present their preliminary results on modularity.

Fourth Week

This week was devoted to the study of the homology of the complement of the complexification of a real arrangement. I first distributed excerpts on basic notions of homological algebra from MacLane's book (1994). We spent time studying these notes, and I lectured on this subject, which was understandably more difficult for them to grasp than the previous material.

Subsequently, they were better prepared to understand the construction of the Orlik-Solomon algebra on lattices. On Wednesday, I spent time with each group introducing their respective reflection groups, D_n, B_n, and F_4. The symmetric group S_n had already been introduced the previous week in connection with the partition lattice. In particular, students were told about the link between the exponents of a reflection group and the factorization of the corresponding Poincaré polynomial. They were then able to find the exponents for their respective groups and so factorized their polynomial.

After this, they learned how to construct a nonbroken circuit basis for the Orlik-Solomon algebra associated to a geometric lattice. The presentations on Friday consisted of a report of their progress on the factorization of the Poincaré polynomial and on the modularity and supersolvability properties of their lattice. Notes were distributed on Wednesday and Friday.

Fifth Week

Class time this week was devoted entirely to the projects; no new concepts were studied. During each class the TA and I spent time with the students helping them to better understand the concepts of homology, and of the Orlik-Solomon algebra and its corresponding nonbroken circuit (NBC) basis. Their projects consisted of determining whether their arrangements were supersolvable, in finding the corresponding Poincaré polynomial together with its factorization, and lastly describing an NBC basis for the Orlik-Solomon algebra.

It should be mentioned that among all reflection arrangements only the S_n, B_n, and dihedral arrangements are supersolvable (Barcelo and Ihrig, 1995).

Sixth Week

The final week was devoted to the presentation of the projects. The students working on the B_n arrangement showed, using very elegant combinatorial arguments, that their arrangement was supersolvable (they exhibited all the possible modular chains of $L(B_n)$). Therefore they could easily calculate the Poincaré polynomial. They also found an NBC basis using a *hand-procedure* similar to the one found in Barcelo and Goupil (1995).

The students working on the arrangement D_n had more difficulties. They showed that none of the el-

ements of rank 2 of $L(D_n)$ was modular, correctly concluding that there was no possible modular chain in $L(D_n)$ and that $L(D_n)$ is not a supersolvable lattice. They were able to calculate the Poincaré polynomial by finding the exponents of the group. They obtained a nice form for the matrix corresponding to a Coxeter element of D_n for any n, thus enabling them to find, using elementary arguments, all of the exponents of D_n. Their difficulties were in trying to characterize an NBC basis. They eventually needed to look at a description of such a basis given in Barcelo and Goupil (1995).

The group working on F_4 gave a spectacular presentation. They showed that F_4 was not supersolvable by showing that there were no modular elements of rank 2. Even though it is known that F_4 is not supersolvable, one previous proof requires very sophisticated tools from algebraic geometry. The only other previously known proof is of a more general nature; it includes all reflection arrangements other than S_n, B_n, and the dihedral ones, and requires more knowledge of group theory. Thus their proof is the first elementary one; it is simple, elegant, and understandable to any undergraduate student with a basic knowledge of linear algebra. They were also able to factorize the Poincaré polynomial by finding the exponents of F_4, and to describe an NBC basis.

First Week

In 1943, J.L. Woodbridge proposed the following problem.

- Show that n cuts can divide a cheese into as many as
$$\frac{(n+1)(n^2-n+6)}{6}$$
pieces.

It is easy to see that n points can divide a line into $1+n$ parts. Similarly, n lines can divide a plane into
$$1+n+\binom{n}{2}$$
parts, and n planes can divide space into
$$1+n+\binom{n}{2}+\binom{n}{3}$$
parts.

In order to maximize the number of pieces, the planes must be in general position, that is, any two planes must have a common line and these two lines must be distinct; any three planes have a common point and these points must be distinct. Allowing degeneracy makes the problem of counting parts much harder. In the following notes we consider arrangements of planes in higher dimensional spaces (hyperplane arrangements). We start with a brief summary of some of the various areas of mathematics involved in the study of hyperplane arrangements.

1. Braid Arrangement (symmetric group).

The *braid arrangement* consists of *hyperplanes*

$$H_{i,j} = \ker(z_i - z_j)$$

in $\mathbb{C}^\ell$. The complement of the union of these hyperplanes,

$$M = \{z \in \mathbb{C}^\ell \mid z_i \neq z_j \text{ for } i \neq j\},$$

is called the *pure braid space*.

The *Poincaré polynomial* $Poin(M,t)$ (due to V. I. Arnold, 1969) of the pure braid space is given by the beautiful formula:

$$Poin(M,t) = (1+t)(1+2t)\dots(1+(\ell-1)t).$$

Arnold also constructed a *graded algebra* A, and showed that there is an isomorphism of graded algebras

$$H^*(M) \simeq A,$$

between the *cohomology ring* of M and A. This construction gives a simple presentation of the cohomology ring of the pure braid space in terms of *generators* and *relations*.

2. Finite Coxeter Groups.

Brieskorn generalized Arnold's results. He replaced the *symmetric group* and the braid arrangement by a finite *Coxeter group* W and its *reflection representation* in a real vector space $V_{\mathbb{R}}$ of dimension l. Now if we let V be the *complexification* of $V_{\mathbb{R}}$, and $M_W \subset V$ be the complement of the reflecting hyperplanes of W, we get that

$$\begin{aligned} Poin(M_W,t) &= (1+m_1t)(1+m_2t)\dots \\ &\quad (1+m_lt), \end{aligned}$$

where the m_i's are certain integers called the *exponents* of W (Brieskorn,1973).

3. Lattice of Intersections of Hyperplanes.

Looking at an arrangement of lines (planes) in $\mathbb{R}^2$ ($\mathbb{R}^3$), one sees that the plane (space) is divided into *chambers*. In order to count the number of chambers of an arbitrary real arrangement A, Zaslavsky (1975) defined the set $L(A)$ of intersections of hyperplanes of A, and *partially ordered* $L(A)$ by reverse set inclusion. He then introduced the method of *deletion* and *restriction* to obtain recursion formulas for counting problems. Using the *Möbius function* of $L(A)$, he defined the *characteristic polynomial* $\Pi(A,t)$ of $L(A)$, and proved that

$$\text{the number of chambers} = \Pi(A,1).$$

Later, Orlik and Solomon (1980) showed that

$$\Pi(A,t) = Poin(M_W,t).$$

Thus the *Betti numbers* of the complement depend only on the lattice of intersections of the hyperplanes!

4. Orlik-Solomon Algebra.

Orlik and Solomon (1980) defined a graded algebra $\mathcal{O}(A)$ which is constructed using only the lattice $L(A)$, and showed that there is an isomorphism of graded algebras

$$H^*(M(A)) \simeq \mathcal{O}(A).$$

In order to study the structure of the algebra $\mathcal{O}(A)$, it is useful to have a standard way to choose a basis. Such a basis, called the *nonbroken circuit* basis, was constructed independently by Björner (1993), Gelfand and Zelevinsky (1986), and Terao and Jambu (1989).

We will study the braid arrangement as well as the B_n, D_n, and F_4 arrangements. In particular we will study their lattices of intersections $L(A)$, their Poincaré polynomials, their nonbroken circuit (NBC) bases, and their Orlik-Solomon algebras. As a team project each group is responsible for finding the Poincaré polynomial associated to their arrangement as well as to construct a nonbroken circuit basis. The Poincaré polynomials can be found in the literature. On the other hand, even though there is a general method to construct NBC bases, it can be quite challenging to actually do so for a particular arrangement. For example, there are several NBC bases for S_n, which can be found in the literature. On the other hand, for B_n and D_n there are only a couple of bases in the literature, and for F_4, to my knowledge, none has been constructed yet.

Exercises

1. Discuss and prove:

a) n points can divide a line into $1 + n$ parts.

b) n lines can divide a plane into

$$1 + n + \binom{n}{2}$$

parts.

c) n planes can divide space into

$$1 + n + \binom{n}{2} + \binom{n}{3}$$

parts.

d) n cuts can divide a cheese into as many as

$$\frac{(n+1)(n^2 - n + 6)}{6}$$

pieces.

2. Discuss and prove:

n lines in general position divide the plane into

$$R = 1 + n + \binom{n}{2}$$

regions of which

$$R' = 1 - n + \binom{n}{2} = \binom{n-1}{2}$$

are bounded.

3. Lines in the plane can fail to be in general position in two different ways. Can you describe these two ways?

4. Both kinds of degeneracies reduce the number of points of intersection. Why?

5. Consider a multiple point M of multiplicity λ. The concurrency of those λ lines at M causes the loss of how many regions?

6. Now consider a family of μ parallel lines in a direction d. Imagine the lines displaced a little to pass through a common point M far away, and rebuild the parallels by letting M tend to infinity in the direction d.

a) How many regions are lost at the point M?

b) How many regions are lost beyond M?

c) So what is the total loss due to the μ parallel lines?

7. From these considerations, if there are m multiple points $M_1, M_2, \ldots, M_m$ in the given arrangement of n lines with, say, $\lambda_i \geq 3$ lines passing through M_i then what ought to be the total loss of regions?

b) If there are p parallel families with, say, $\mu_j \geq 2$ lines in the jth family, then the total loss of regions due to the parallels ought to be what?

c) Hence, what is the formula for the total number of regions formed by n lines?

8. This was Roberts's heuristic argument; it is certainly convincing, not to say compelling! Try the formula for a few non-trivial cases. Does it work?

9. These arguments do not immediately yield a correct proof. Can you see what could go wrong?

10. Try proving it.

Second Week: A

Let K be a *field*, let V_K be a *vector space* over K. A **hyperplane** H is a *codimension* 1 (affine) *subspace* of V_K. A **hyperplane arrangement** A_K is a finite set of hyperplanes in V_K. Note that for our purpose K will be $\mathbb{R}$ unless stated otherwise, and we shall write V and A when there is no danger of confusion. If the dimension of V is n, we say that A is an n-arrangement.

Each hyperplane is the *kernel* of a polynomial α_H of degree 1 defined up to a constant (in d variables, if $d = dim V_K$).

The product

$$Q(A) = \prod_{H \in A} \alpha_H$$

is called the **defining polynomial** of A. By convention, $Q(\emptyset_l) = 1$ is the defining polynomial of the empty arrangement.

A is **centerless** if

$$\bigcap_{H \in A} = \emptyset.$$

If

$$T = \bigcap_{H \in A} H \neq \emptyset,$$

then A is a **centered arrangement** with **center** T. If A is centered, then coordinates may be chosen so that each hyperplane contains the origin. In that case we say that A is **central**.

Examples of real arrangements.

Let $V_\mathbb{R}$ be the one-dimensional *Euclidean space*, that is: _____. There is (are) _____ central arrangement(s) which consist(s) of: _____. A real affine 1-arrangement consists of: _____. A central real 2-arrangement is: _____. An affine 2-arrangement is: _____.

Exercises

1. Given the following central 2-arrangement, find its defining polynomial.

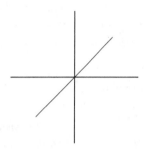

2. Given the following affine 2-arrangement, find its defining polynomial.

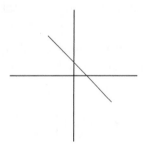

3. Define A by

$$Q(A) = y(x + 2y).$$

Then A is a _____ arrangement and consists of _____. Draw A in the plane $\mathbb{R}^2$.

4. Define A by

$$Q(A) = x(x + 2)y(y - 2)(x + y).$$

Then A is a _____ arrangement and consists of _____ lines. Again draw this arrangement A.

5. Consider the cube in $\mathbb{R}^3$ with vertices $(\pm 1, \pm 1, \pm 1)$. The 9 planes of symmetry form a central 3-arrangement defined by $Q(A) = $ _____. Can you draw this arrangement?

6. The arrangement of the coordinate hyperplanes in $\mathbb{R}^n$ is defined by $Q(A) = $ _____. We will see later that this arrangement is isomorphic to the Boolean lattice; thus, it is called the Boolean arrangement.

7. For $1 \leq i < j \leq n$, where $V = \mathbb{R}^n$, let $H_{i,j} = \ker(x_i - x_j)$. The braid arrangement (or A_{n-1}-arrangement) is defined by

$$Q(A) = \prod_{1 \leq i < j \leq n} (x_i - x_j).$$

Describe, in terms of subspaces, the hyperplanes of the braid arrangement for $n = 3, 4$.

8. a) For $1 \leq i < j \leq n$, let

$$\begin{aligned} H_{i,j} &= \ker(x_i - x_j), \\ \overline{H}_{i,j} &= \ker(x_i + x_j), \\ H_i &= \ker(x_i). \end{aligned}$$

The set of all these hyperplanes forms the hyperoctohedral arrangement (or the B_n-arrangement). The defining polynomial of B_n is: _____.

b) Describe, in terms of subspaces, the hyperplanes of the hyperoctohedral arrangement.

9. For $1 \leq i < j \leq n$, consider the set of hyperplanes $H_{i,j}$ and $\overline{H}_{i,j}$. This also constitutes an arrangement whose defining polynomial is $Q(A) = $ _____. This arrangement is called the D_n-arrangement.

10. Let $V = \mathbb{R}^4$. For $1 \leq i < j \leq 4$, consider the set of hyperplanes $H_{i,j}, \overline{H}_{i,j}, H_i$ and the 8 hyperplanes defined by $\ker(x_1 \pm x_2 \pm x_3 \pm x_4)$. This is the F_4-arrangement with defining polynomial: _____.

A Combinatorial Tool: $L(A)$.

We now define a most important tool in the study of hyperplane arrangements: a poset $L(A)$ associated with each arrangement. A poset P is a set with a binary relation $\leq$ satisfying the following 3 properties for all $x, y, z \in P$,

1. Reflexivity: _____.

2. Antisymmetry: _____.

3. Transitivity: _____.

A set P together with a binary relation satisfying these three properties is a partially ordered set, or a **poset**. A fourth axiom would be needed to make it a total order. What is this axiom? _____.

We say that a **covers** b in a poset P if $a > b$, but that $a > x > b$ for NO $x \in P$.

Let A be an arrangement and let $L(A)$ be the set of all the nonempty intersections of elements of A. Define a partial order on $L(A)$ by $X \leq Y \Leftrightarrow Y \subseteq X$. Thus $L(A)$ is a partially ordered set (poset).

Exercises

1. Draw the Hasse diagram of the arrangements of Exercises 1-5, and of the Boolean arrangement for $n = 2, 3, 4$. Do you see a pattern for the Hasse diagram of the Boolean arrangement in general?

2. Draw the Hasse diagram of the braid arrangement for $n = 3, 4$. Do you see a way of describing the Hasse diagram of the braid arrangement in general?

3. Draw the Hasse diagram of the B_n-arrangement for $n = 3, (4?)$. Any general picture?

4. Draw the Hasse diagram of the D_n-arrangement for $n = 3, (4?)$. Any general picture?

5. Draw (as much as you can!) the Hasse diagram of the F_4-arrangement.

This notion of poset associated with an arrangement sheds a new light on the problem of isomorphism of arrangements, which we shall revisit before pursuing their study.

Second Week: B

Take a finite set of hyperplanes of a d-dimensional Euclidean space. When these hyperplanes are removed, the remainder of the space decomposes into components, each one a d-dimensional open polyhedron (not necessarily bounded). The original set of hyperplanes, together with the set of all the k-dimensional open faces of all these polyhedra for $-1 \leq k \leq d$, is known as a **cell-complex decomposition** of the space.

Let the Euclidean space be $\mathbb{R}^2$, and consider a finite set of hyperplanes – that is, a finite set of _____. Then the 2-dimensional faces are _____; the 1-

dimensional faces are _____; the 0-dimensional faces are _____; the 1-dimensional face is _____.

An arrangement is called **simple** if its hyperplanes are in general position. In that case, the number of regions (also called **facets**, that is the d-dimensional faces), as well as the number of k-faces and bounded faces have been computed by many authors for Euclidean and projective spaces of various dimensions. Do you know some of these results?

The dimension of the space and the number of hyperplanes do NOT determine the combinatorial type of an arrangement. Nevertheless, Zaslavsky (1975) showed that using only $L(A)$, one could calculate the number of regions determined by the arrangement A.

Definition 1. *Two arrangements are combinatorially equivalent, or isomorphic, if there is a one-to-one correspondence of their faces preserving all incidences. This is an isomorphism of the cell-complex decomposition of the space induced by the arrangement.*

Exercises

Consider the two arrangements of the figure below.

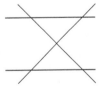

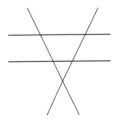

1. For each of these arrangements describe the cell-decomposition. This means find the 0-dimensional faces, 1-dimensional faces and the 2-dimensional faces.

2. For each of the facets write down the number of 1-dimensional faces incident to the given facet. Are these arrangements isomorphic?

A weaker notion is geometric equivalence, which means that the linear relations (and parallelisms) of the hyperplanes are the same.

Definition 2. *We say that two arrangements A and B are geometrically equivalent if their associated lattices $L(A)$ and $L(B)$ are isomorphic.*

Lemma 1. *If $A \simeq B$ (combinatorially equivalent) then $L(A) \simeq L(B)$.*

Exercises

1. Prove this lemma.

2. Is the converse true? Prove or give a counterexample.

3. Construct and discuss the lattices associated to the arrangements given in the figure below.

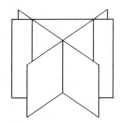

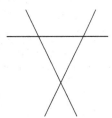

Let us now get back to the study of the poset $L(A)$, together with some of its properties.

Definition 3. *A rank function on $L(A)$ is defined by setting $r(X) = codim(X) \; \forall X \in L(A)$, where*

$codim(X)$ *represents the codimension of the vector space X.*

Thus $r(V) =$ _____; and when H is a hyperplane of A, $r(H) =$ _____.

The elements of rank 1 are called the atoms of $L(A)$. Thus the set of all atoms of A consist of _____.

Next we define two operations on $L(A)$.

Definition 4. *Let $X, Y \in L(A)$. The meet of X and Y, denoted by $X \wedge Y$, is given by*

$$X \wedge Y = \cap \{Z \in L(A) \mid X \cup Y \subseteq Z\}.$$

If $X \cap Y \neq \emptyset$ then denote the join of X and Y by $X \vee Y = X \cap Y$.

Familiarize yourself with these notions using the posets A_{n-1}, B_n, D_n, and F_4.

The notions of rank, meet, and join can be defined on any poset. Let S be any nonempty subset of P; an element $a \in S$ is said to be **minimal** if there is no $x \in S$ such that $a > x$; an element $a \in S$ is said to be **maximal** if there is no $x \in S$ such that $a < x$. A **chain** C is a poset in which any two elements are comparable. The **length** $l(C)$ of a finite chain C is defined by $l(C) = |C| - 1$. If every maximal chain of P has the same length n, then P is said to be **graded** of **rank** n. In this case, there is a unique **rank function** $r : P \to \{0, 1, \ldots, n\}$ such that $r(x) = 0$ if x is a minimal element of P, and $r(y) = r(x) + 1$ if y covers x. If $r(x) = i$, then we say that x has **rank** (height) i.

If X and Y belong to P, then an **upper bound** of X and Y is an element $Z \in P$ such that $Z \geq X$ and $Z \geq Y$. A **least upper bound** of X and Y is an upper bound Z of X and Y such that every upper bound W of X and Y satisfies $W \geq Z$. If a least upper bound of X and Y exists, then it is (clearly) unique and is denoted by $X \vee Y$ (X join Y).

Similarly one can define the lower bound of X and Y. A greatest lower bound of X and Y is a _____. If a greatest lower bound of X and Y exists, then it is denoted by _____ (X meet Y).

A **lattice** is a poset L such that for all pairs of elements $X, Y \in L$, both $X \vee Y$ and $X \wedge Y$ exist. Clearly all finite lattices have a $\hat{0}$ (minimal element) and a $\hat{1}$ (maximal element).

Exercises

1. Using a few of your own posets, give examples and counterexamples of each of the previous definitions.

2. Verify that the operations $\wedge$ and $\vee$ of Definition **4** are indeed meet and join operations as defined above. Do the same for the rank function of Definition **3**.

In order to better comprehend the role the lattice $L(A)$ is playing in the study of arrangements, we need a few more definitions.

Definition 5. *A finite lattice L is* **modular** *iff it is graded and its rank function r satisfies:*

$$r(x) + r(y) = r(x \wedge y) + r(x \vee y).$$

L is said to be **semimodular** *if*

$$r(x) + r(y) \geq r(x \wedge y) + r(x \vee y).$$

A lattice is said to be **atomic** *if every element (except $\hat{0}$) of L is a join of atoms.*

A finite semimodular atomic lattice is said to be a finite **geometric lattice**.

Sometimes it is handy to have a top element in $L(A)$, so we add one denoted by $\hat{1}$. Let $\hat{L}(A) = L(A)$ if A is central, and $\hat{L}(A) = L(A) \cup \hat{1}$ if A is non-central.

Again familiarize yourself with these notions using the posets A_{n-1}, B_n, D_n, and F_4.

Proposition 2. *(see Orlik and Terao, p. 24) Let A be an arrangement and let $L = L(A)$.*

1. *Every element of $L \setminus \{V\}$ is a join of atoms.*

2. *For every $X \in L$, all unrefinable linearly ordered subsets*

   $$V = X_0 < X_1 < \ldots < X_p = X$$

   have the same cardinality. Thus $L(A)$ is a geometric poset.

3. *If A is central, then all joins exist, so L is a lattice. For all $X, Y \in L$ the rank function satisfies*

   $$r(x) + r(y) \geq r(x \vee y) + r(x \wedge y).$$

   Thus for a central arrangement, $L(A)$ is a geometric lattice.

Exercises

1. Prove the previous proposition.

2. Can you give an example of a poset that does not satisfy condition 2 of the above proposition?

We now define a crucial function on posets. The **Möbius function** is defined recursively on any finite poset by:

$$
\begin{aligned}
\mu(x,x) &= 1 \\
\mu(x,y) &= -\sum_{x \leq z < y} \mu(x,z) \text{ if } x < y \\
\mu(x,y) &= 0 \text{ if } x \not\leq y.
\end{aligned}
$$

One can show that in a geometric lattice the sign of $\mu(s,t)$ is $(-1)^{r(t)-r(s)}$. To every finite poset there is an associated polynomial, the **characteristic polynomial** given by:

$$\chi_{\hat{L}(A)}(t) = \sum_{X \in \hat{L}(A)} \mu(\hat{o}, X) t^{r(\hat{1}) - r(X)}.$$

As we shall see in the next theorem, the characteristic polynomial is crucial for, when evaluated at $t = -1$, it yields the total number of regions in the arrangement.

Theorem 3. *Let A be an Euclidean arrangement of hyperplanes. The total number of regions $C(A)$ is:*

$$C(A) = \sum_{x \in L(A)} |\mu(\hat{o}, x)| = (-1)^{r(A)} \chi_{L(A)}(-1).$$

This theorem will be proved by one of the groups during our next Friday meeting. The proof relies on the notions of induced arrangements and upper semi-lattices. Let S be any subspace of the whole space $\mathbb{R}^d$. The arrangement A **induces** on S an arrangement

$$A_S = \{H \cap S : H \in A, \dim(H \cap S) = \dim S - 1\}.$$

Let X be an element of A. The upper semi-lattice with respect to X is

$$USL(X) = \{Y \in L(A) : Y \geq X\}.$$

Exercises

1. Calculate the characteristic polynomial of the arrangements 1-7 given at the beginning of this week's handout.

2. What is the link between $L(A_S)$ and $USL(X)$?

3. Prove that $r(A_S) = r(A) - r(S)$.

4. How is the characteristic polynomial of A_S related to the one of $USL(X)$?

5. In this exercise you are asked to prove the Möbius inversion formula. Let f, g be functions of $L(A)$ with values in $\mathbb{C}$. Then $\forall y \in P$

$$g(Y) = \sum_{X \leq Y} f(X)$$

$$\Leftrightarrow \quad f(Y) = \sum_{X \leq Y} g(X) \mu(X, Y).$$

a) Step 1: Prove

$$\sum_{X \leq Z \leq Y} \mu(Z, Y) = 0$$

if $X, Y \in L$ and $X < Y$.

Hint: Write $L(A) = \{X_1, \ldots, X_r\}$ where the numbering is chosen so that $X_i \leq X_j \Rightarrow i \leq j$. Let A be the $r \times r$ matrix whose (i, j) entry is $\mu(X_i, X_j)$. Let B be the $r \times r$ matrix whose (i, j) entry is 1 if $X_i \leq X_j$, 0 otherwise.

b) What can you say about the "aspect" of A and B?

c) What is AB equal to?

d) Thus, BA is equal to what?

e) Conclude that

$$\sum_{X \leq Z \leq Y} \mu(Z, Y) = 0$$

if $X, Y \in L$ and $X < Y$.

f) Step 2:

$$\sum_{Z \leq Y} \mu(Z, Y) g(Z) = \sum_{Z \leq Y} \mu(Z, Y) \sum_{X \leq Z} f(x)$$

$$= \sum_{X \leq Y} \sum_{X \leq Z \leq Y} \mu(Z, Y)) f(x)$$

$$= \quad ?$$

Third Week

The characteristic polynomial associated to the lattice of intersections of an $\ell-$arrangement of hyperplanes is one of the most important combinatorial invariants. Another closely related polynomial is the Poincaré polynomial of A:

$$\Pi(A, t) = \sum_{X \in L(A)} \mu(\hat{o}, X)(-t)^{r(X)}.$$

You can verify that the relation between the characteristic polynomial $\chi(A, t)$ and the Poincaré polynomial is:

$$\chi(A, t) = t^\ell \Pi(A, -t^{-1}).$$

An interesting aspect of the Poincaré polynomial is its factorization into linear factors, with integer coefficients, for some classes of arrangements. One of these classes is the class of reflection arrangements. A reflection arrangement is obtained from a (real) reflection group. The braid arrangement and the B_n, D_n and the F_4 arrangements are all reflection arrangements. We will see later to which reflection groups they are associated. Among those arrangements for which the Poincaré polynomial factors are the supersolvable ones. These arrangements are characterized by the fact that their lattices $L(A)$ contain a chain of modular elements. When an arrangement is supersolvable, it is relatively easy to find its factorization.

Let L be any lattice. A pair $(X, Y) \in L \times L$ is called a **modular pair** if for all $Z \leq Y$

$$Z \vee (X \wedge Y) = (Z \vee X) \wedge Y.$$

It is possible to show that a pair $(X, Y) \in L \times L$ is modular if and only if

$$r(X) + r(Y) = r(X \vee Y) + r(X \wedge Y).$$

An element $X \in L$ is said to be **modular** if (X, Y) is a modular pair for all $Y \in L$. Modular elements in a lattice can be characterized by the following criteria: an element $X \in L$ is modular if and only if (X, Y) is a modular pair for every $Y \in L$ such that $X \wedge Y = V$.

Let A be an arrangement with $r(L(A)) = \ell$. A is said to be **supersolvable** if $L(A)$ has a maximal chain of modular elements. As mentioned earlier, the advantage for a lattice being supersolvable resides in the fact that its Poincaré polynomial is very easy to calculate.

Theorem 4. *(see Orlik and Terao, p. 49) Let A be a supersolvable ℓ-arrangement with a maximal unrefinable chain of modular elements,*

$$V = X_0 < X_1 < \ldots < X_\ell = T.$$

Let $b_i = |A_{X_i} \setminus A_{X_{i-1}}|$. *Then*

$$\Pi(A, t) = \prod_{i=1}^{\ell} (1 + b_i t).$$

The proof of this theorem can be done in three steps, each of them requiring you to prove a given claim. This is done at the end of the exercises below, following Orlik and Terao's proof (p. 69).

Exercises

1. Calculate the Poincaré polynomial of the Boolean arrangement.

2. Find modular elements in the Boolean lattice.

3. Find modular elements for the braid arrangements.

4. Start looking for modular elements in the arrangement you chose for your project.

5. Is the Boolean arrangement supersolvable? If yes, exhibit at least one maximal chain of modular elements.

6. Same question for the braid arrangement.

7. Is the arrangement you chose for your project supersolvable?

8. Calculate the Poincaré polynomial for the braid arrangement.

9. Can you calculate the Poincaré polynomial of your arrangement?

10. We shall now prove the previous theorem by induction on the number of hyperplanes in A.

Let $|A| = 1$. We have that $V = X_0 < X_1 = T$, thus

$$\prod_{i=1}^{1} (1 + b_i t) = \underline{\hspace{2cm}}$$

while

$$\prod(A, t) = \underline{\hspace{2cm}}.$$

Next assume the theorem true for the arrangement A when $|A| \leq n - 1$.

Let H_0 be an atom not below $X_{\ell-1}$. So,

$$X_{\ell-1} \vee H = X_\ell = T.$$

We wish to use the deletion-restriction theorem. For this, we will consider two arrangements, both of which have their number of atoms $\leq (n-1)$.

i) $A' \equiv A \setminus \{H_0\}$.

ii) $A'' = \{H_0 \cap H \mid H \in A \text{ and } H \neq H_0\}$.

Notice that A' has $(n-1)$ atoms.

What is the maximum number of hyperplanes in A''? Thus $|A''| < |A|$.

Note that $A_{X_{\ell-1}}$ is the set of all hyperplanes below $X_{\ell-1}$, and as such $L(A_{X_{\ell-1}})$ is supersolvable, since the chain

$$V = X_0 < X_1 < \ldots < X_{\ell-1}$$

belongs to $L(A_{X_{\ell-1}})$ and is certainly modular in $L(A_{X_{\ell-1}})$.

Claim 1: $L(A'')$ is isomorphic to $L(A_{X_{\ell-1}})$ with modular chain given by

$$
\begin{aligned}
H_0 &= X_0 \vee H_0 < X_1 \vee H_0 \\
&= Y_1 < X_2 \vee H_0 \\
&= Y_2 < \ldots < X_{\ell-1} \vee H_0 \\
&= Y_{\ell-1}.
\end{aligned}
$$

Recall that H_0 was chosen so that $H_0 \not< X_{\ell-1}$, which implies that $H_0 \not< X_i$, $\forall i = 1, \ldots, \ell - 1$.

Claim 2: $|A''_{Y_i} \setminus A''_{Y_{i-1}}| = |A_{X_i} \setminus A_{X_{i-1}}|$. Thus by induction

$$\prod(A'', t) = \prod_{i=1}^{\ell-1} (1 + b_i t).$$

Claim 3: A' is supersolvable. Thus we can apply the induction hypothesis on A'. For this we need to find a modular chain in A'. This depends on H_0. A hyperplane H_0 is said to be a separator if the maximal element, $T(A)$, of $L(A)$ does not belong to $L(A')$. We consider two cases:

i) H_0 is not a separator, i.e. $T(A) \in L(A')$.

In this case

$$V = X_0 < X_1 < \ldots < X_\ell = T(A)$$

is a maximal chain of modular elements in $L(A')$. Note that the number of hyperplanes of $A' = A \setminus H_0$ below X_i but not below X_{i-1} is given by $b_i, \forall i = 1, \ldots, \ell - 1$, and

$$|A'_{X_\ell} \setminus A'_{X_{\ell-1}}| = b_\ell - 1.$$

So

$$\prod(A',t) = \prod_{i=1}^{\ell-1}(1+b_i t)(1+(b_\ell-1)t).$$

ii) H_0 is a separator, i.e. $T(A) \notin L(A')$.

This implies that

$$V = X_0 < X - 1 < \ldots < X_{\ell-1}$$

is a modular chain of elements in A', and

$$|A_{X_\ell} \setminus A_{X_{\ell-1}}| = 1,$$

i.e. $b_\ell = 1$. Thus,

$$\prod(A',t) = \prod_{i=1}^{\ell-1}(1+b_i t).$$

a) Follow this proof using the braid arrangement and the maximal modular chain

$$V = X_0 < 12 < 123 < \ldots < \hat{1}.$$

b) Prove the 3 claims, and complete the proof of the theorem.

Fourth Week: A

Let A be a real ℓ-arrangement with defining polynomial $Q(A)$. The complexification of a real arrangement is an arrangement $A_{\mathbb{C}}$ of hyperplanes in $\mathbb{C}^\ell$ with the same defining polynomial $Q(A_{\mathbb{C}}) = Q(A)$. This time though, the kernel of the defining polynomial takes its values in $\mathbb{C}^\ell$. In general, it is quite difficult to visualize these complexifications. For example, if $Q(A) = xy(x+y)$ in real dimension, it consists of three 2-planes in 4-space which meet at the origin.

The complement $M(A)$ of an arrangement is given by

$$M(A) = \mathbb{C}^\ell \setminus Q(A_{\mathbb{C}}),$$

where $Q(A_{\mathbb{C}})$ is the union of all complexified hyperplanes in A. As we mentioned earlier, we shall study the homology groups of $M(A)$ for different arrangements. For this we proceed as follows:

1. We will define the Orlik-Solomon algebra $\mathcal{O}(A)$ over the lattice $L(A)$. This algebra is isomorphic to the homology algebra of $M(A)$. In particular, this means that the dimensions of ith-homology, for each i, are the same. Since the Poincaré polynomial is the enumerating polynomial for these dimensions, that is

$$Poin(M(A),t) = \sum_{i \geq 0} dim(H^i(M(A))t^i,$$

we have that:

$$Poin(\mathcal{O}(A),t) = Poin(M(A),t).$$

More importantly there is a crucial theorem that says that $Poin(\mathcal{O}(A),t) = \Pi(A,t)$.

2. Thus in order to calculate the Poincaré polynomial of an arrangement, it will suffice to find a basis for each of the components of $\mathcal{O}(A)$. We shall study the nonbroken circuit basis (NBC-basis) of $\mathcal{O}(A)$.

Let us start with the definition of an algebra. An algebra $\mathcal{A}$ is a vector space (over some field F) endowed with a bilinear operation: $\diamond : \mathcal{A} \times \mathcal{A} \to \mathcal{A}$. Equivalently, one may define an algebra as a ring $(\mathcal{A}, \diamond, +)$ together with an action of the field F on $\mathcal{A}$ (the scalar multiplication) which is compatible with the multiplication "$\diamond$" and addition "$+$".

In our situation, the field F should be $\mathbb{R}$ unless otherwise stated. We shall give a definition of the Orlik-Solomon algebra for central arrangements that follows the one given by Orlik and Terao (Chapter 3, Section 1). An equivalent definition exists for affine arrangements (see Orlik and Terao, 1992). Hereafter all arrangements will be **central**.

Definition 6. *Let A be a real (central) arrangement. Let E_1 be the vector space over $\mathbb{R}$ spanned by the set $\{e_H\}_{H \in A}$, i.e. $E_1 = \bigoplus_{H \in A} \mathbb{R} e_H$. (The dimension of $E_1 = |A|$.) Let $E = \wedge(E_1)$ be the exterior algebra of E_1, and for any elements $e_H, e_{H'} \in E_1$ denote the bilinear operator $\diamond$ by: $e_H \diamond e_{H'} = e_H e_{H'}$.*

The exterior algebra E can be viewed as the algebra whose elements are concatenations of elements of E_1 subject to the following rules:

R1: $e_H e_H = 0$,

R2: $e_H e_K = -e_K e_H$.

For $p \geq 2$, let E_p be the linear span (over $\mathbb{R}$) of all elements of the form: $e_{H_1} e_{H_2} \ldots e_{H_p}$ with $H_k \in A$. Let $E_0 = \mathbb{R}$, and let E_1 be as defined earlier. Thus if $|A| = n$ then $E = \bigoplus_{p=0}^{n} E_p$.

Definition 7. *Define a linear map $\partial : E \to E$ by*

i) $\partial 1 = 0$

ii) $\partial e_H = 1$

iii) *for* $p \geq 2$,

$$\partial(e_{H_1} e_{H_2} \cdots e_{H_p})$$
$$= \sum_{k=1}^{p} (-1)^{k-1} e_{H_1} \cdots \hat{e}_{H_k} \cdots e_{H_p}$$

for all $H_1, \ldots, H_p \in A$, *where* $\hat{e}_{H_k}$ *means that* e_{H_k} *is deleted from* $e_{H_1} \cdots e_{H_k} \cdots e_{H_p}$.

Given a p-tuple of hyperplanes, $S = (H_1, \ldots, H_p)$, write $|S| = p$, $e_S = e_{H_1} \cdots e_{H_p} \in E$, and $\cap S = H_1 \cap \ldots \cap H_p$. Since A is central, $\cap S \in L(A)$, $\forall S$. If $p = 0$ we let $S = ()$ be the empty-tuple and $e_S = 1$ and $\cap S = V$. Also note that:

$$r(\cap S) = r(\vee S) \leq |S|.$$

Definition 8. *A set* S *of hyperplanes is said to be independent if* $r(\cap S) = |S|$, *and it is said to be dependent if* $r(\cap S) < |S|$.

In particular, this means that the tuple S is independent if the corresponding linear forms $\alpha_1, \ldots, \alpha_p$ are linearly independent. Equivalently, the hyperplanes of S are in general position.

Let S_p denote the set of all p-tuples $(H_1, \ldots, H_p)$ of hyperplanes, and let $\mathcal{S} = \bigcup_{p \geq 0} S_p$.

Definition 9. *Let* A *be an arrangement. Let* $I = I(A)$ *be the ideal of* E *generated by* ∂e_S *for all dependent* $S \in \mathcal{S}$. *Let* $I_p = I \cap E_p$. *Then,*

$$I = \bigoplus_{p=0}^{n} I_p.$$

Finally, we give the definition of the Orlik-Solomon algebra associated with an arrangement.

Definition 10. *Let* A *be an arrangement. The Orlik-Solomon algebra associated to* A *is given by* $\mathcal{O}(A) = E \setminus I$.

Let $\varphi : E \to E \setminus I$ be the natural homomorphism $\varphi(E_p) = [E_p]$. Denote $\varphi(e_H) = a_H$, and if $S \in \mathcal{S}$, $\varphi(e_S) = a_S$.

Since both E and I are graded, $\mathcal{O}(A)$ is a graded anticommutative algebra. Since the elements of S_1

are independent, we have that $I_0 = 0$ and $\mathcal{O}_0 = E_0 \setminus I_0 = \mathbb{R}$. In S_2 the only dependent elements are of the form $S = (H, H)$. But then $e_S = e_H e_H = 0$. Thus $I_1 = 0$, the elements a_H are linearly independent over $\mathbb{R}$, and $\mathcal{O}_1 = \bigoplus_{H \in A} \mathbb{R} a_H$. Therefore,

$$\mathcal{O}(A) = \bigoplus_{p=0}^{l} \mathcal{O}_p.$$

Fourth Week: B

In order to define the nonbroken circuit basis (NBC-basis) for the Orlik-Solomon algebra $\mathcal{O}(A)$ of an arrangement, we first need to choose a total order on the set of atoms of A (i.e. on the set of hyperplanes). Let this order on A be denoted by $<$. Hereafter, an ordered subset of A of cardinality p will be denoted by $(H_1, \ldots, H_p)$ with the understanding that $H_1 < H_2 < \ldots < H_p$.

Definition 11. *A subset of hyperplanes*

$$S = \{H_1, \ldots, H_m\}$$

is said to be a circuit if it is minimally dependent. That is, $S = \{H_1, \ldots, H_m\}$ *is a dependent set*

$$(r(\vee_{i=1}^{m} H_i) < m),$$

but $S \setminus H_i$ *is independent for all* $1 \leq i \leq m$.

Definition 12. *A broken circuit is an ordered tuple* $S = (H_1, \ldots, H_p)$ *such that there exists a hyperplane* $H \in A$ *with* $H < H_i$, $\forall 1 \leq i \leq p$, *and such that* $S \cup H$ *is a circuit.*

We note that every broken circuit is obtained by deleting the minimal element of a circuit.

Definition 13. *A nonbroken circuit set of hyperplanes is an ordered tuple* $(H_1, \ldots, H_p)$ *that does not contain any broken circuit.*

Definition 14. *The NBC complex* $C(A)$ *is defined as follows. Let* K *be any ring (for our purposes* $K = \mathbb{R}$). *Let* $C_0 = K$, *and for* $p \geq 1$, *let* C_p *be the free* K-*module with basis* $\{e_S \in E \mid S \text{ is an NBC element}, |S| = p\}$. *The free graded* K-*module* $C(A)$ *is given by:*

$$C(A) = \bigoplus_{p \geq 0} C_p.$$

Note that by definition $C(A)$ is a submodule of $E(A)$ if $K = \mathbb{R}$. In general, $C(A)$ is not closed under multiplication in $E(A)$, so $C(A)$ is not a subalgebra of $E(A)$.

Recall the natural projection $\varphi : E(A) \to \mathcal{O}(A)$, and let $\psi : C(A) \to \mathcal{O}(A)$ be its restriction. We have the following crucial theorem, which we shall use without proving it.

Theorem 5. *(see Orlik and Terao, 1992) ψ is an isomorphism of graded modules; thus the set*

$$\{e_S + I \in A(A) \mid S \text{ a NBC element }\}$$

is a basis, the NBC-basis, for $\mathcal{O}(A)$ as a graded K-module.

Exercises

1. Let A be defined by

$$Q(A) = xyz(x + y)(x + y - z).$$

Let

$$
\begin{aligned}
H_0 &= \ker(x + y - z), \\
H_1 &= \ker(x), \\
H_2 &= \ker(y), \\
H_3 &= \ker(z), \\
H_4 &= \ker(x + y).
\end{aligned}
$$

Define the linear order on A by $H_i < H_j \Leftrightarrow i < j$. Find the NBC basis for $C(A)$.

2. Let A be the braid arrangement with its hyperplanes denoted by $H_{j,i}$ $(j > i)$, i.e.

$$H_{21}, H_{31}, H_{32}, \ldots.$$

Let the linear order on A be the lexicographic order on the pairs (j, i), where $j > i$. With this order, find the NBC-basis for A_3, A_4. Can you generalize to A_n? For a description of these bases, see Barcelo (1990).

3. Find the NBC-bases for your arrangements.

References

Arnold, V. I. (1969). The cohomology ring of the colored braid group. *Math. Notes* **5**:138–140.

Barcelo, H. (1990). On the action of the symmetric group on the free Lie algebra and the partition lattice. *J. of Comb. Theory (A)* **55**:93–129.

Barcelo, H., and Goupil, A. (1995). Non-broken circuits of reflection groups and factorization in D_n. *Israel J. of Math.* **91**:285–306.

Barcelo, H., and Ihrig, E. (1995). Lattices of parabolic subgroups in connection with hyperplane arrangements. *Preprint.*

Björner, A., Las Vergnas, M., Sturmfels, B., White, N., and Ziegler, G. (1993). Oriented matroids. *Encyclopedia of Mathematics* **46**. Cambridge University Press, Cambridge.

Brieskorn, E. (1973). Sur les groupes de tresses. *Séminaire Bourbaki 1971/72, Lecture Notes in Math.* **317**:21–44. Springer-Verlag, Berlin.

Gelfand, I. M., and Zelevinsky, A. V. (1986). Algebraic and combinatorial aspects of the general theory of hypergeometric functions. *Funct. Anal. and Appl.* **20**:183–197.

Jambu, M., and Terao, H. (1989). Arrangements of hyperplanes and broken circuits. *Amer. Math. Soc.* **90**:147–162.

Maclane, S. (1994). *Homology.* Springer-Verlag, Classics in Math., New York.

Orlik, P., and Solomon, L. (1980). Combinatorics and topology of complements of hyperplanes. *Invent. Math.* **56**:167–189.

Orlik, P., and Terao, H. (1992). *Arrangements of hyperplanes.* Series of Comprehensive studies in Math. **300**. Springer-Verlag, New York.

Roberts, S. (1889). On the figures formed by the intercepts of a system of straight lines in a plane, and on analogous relations in space of three dimensions. *Proc. London Math. Soc.* **19**:402–423.

Wetzel, H. (1978). On the division of the plane by lines. *Amer. Math. Monthly* **85**:648–656.

Zaslavsky, T. (1975). *Facing up to arrangements.* Memoirs of the American Mathematical Society **154**.

Ziegler, G. (1992). *Combinatorial Models for subspace Arrangements.* Habilitations-Schrift, TU Berlin.

Algebra Seminar Taught by a Number Theorist

ANTONIA BLUHER

Stanford University

Introduction

I would like to report on my experience teaching an algebra seminar in the Mills Summer Mathematics Institute. I had seen the students' files in advance and knew this group would be ready for a challenge. They were the type of students who give back more energy than they absorb, and they were excited about heading into research. At the same time, many were worried about how they would integrate present or future family lives with their careers. Several of the women told me they had never had a female professor and had very few, if any, viable role models. Some asked for appointments with me so that they could speak one on one about some of these issues; others talked over lunch. One woman had been told by a male professor in her home institution that she should not marry, as this would distract her too much from her career. The two female faculty in her department were older women with no children, one of whom had never married. This student was amazed by the fact that all four of the instructors at the Mills Summer Mathematics Institute had small children. Another student had been told by a kindly physics professor that he wanted women to work together in a group because they had more difficulty conceptualizing things than men. This woman's exposure to the Summer Mathematics Institute was her first chance to see a different viewpoint.

The First Exercise

The algebra seminar was organized around half-hour lectures followed by a full hour of problem-solving on the board in small groups. This method was effective in teaching a group of students with heterogeneous backgrounds. Everyone got involved, and everyone was thinking, writing, and proving. Apparently many former instructors at the Mills program have discovered this method independently.

We started the seminar with the following board game for two players:

1. Draw a circular game board with 45 squares (see figure).

2. Label one square START, and label an adjacent square END.

3. Each player is given an integer between 1 and 44.

4. The players take turns moving their given number of squares.

5. The winner is the first person who reaches the END square.

For example, suppose Player A has the number 7. Player B has the number 14. The first six moves are shown below as A1, A2, ..., A6 and B1, B2, ..., B6. In this example, Player A will reach END after 13 moves and Player B will reach END after 29 moves, so Player A wins.

Since people get tired of counting, they start to find shortcuts. For instance, six of A's moves or three of B's moves is equivalent to going three squares backwards. This shortcut is related to the Euclidean algorithm. One pair of students noticed that if the last digit of the player's number is a 2 or 7, then the number of moves required to reach END ends in the digit 3 or 8 (provided they can finish). These students had found a relation in mod 5 arithmetic.

Now suppose we give Player A the number 6. After a while, she realizes she will never reach the finish. In fact, after 15 moves she is back on START.

The question arises: How can you tell whether you will reach END? The students arrived at the conjecture: you will reach END if and only if your number is not divisible by 3 or 5. Later, as we delved into ring theory, the class was asked to prove that the numbers which reach END are the units of the ring of integers mod 45, and the numbers which don't finish are the zero-divisors. It was easy to show that zero-divisors don't finish, numbers which finish are units, and that all units will finish. The hard part is to show that every number which doesn't reach END is a zero-divisor. One student's solution was a counting argument which demonstrates that all nonzero elements of a finite commutative ring are zero-divisors or units.

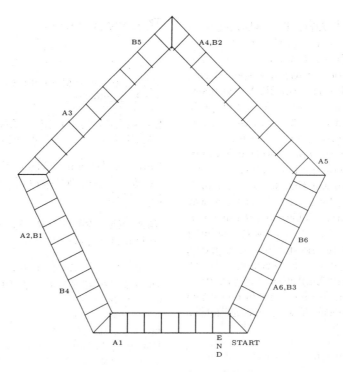

The Seminar

A course by the name of algebra, taught by a number theorist, could go in a myriad of directions. The themes I decided to emphasize were: quadratic reciprocity and the Hasse principle. The material was divided into the following four sections:

I. First we defined rings and studied linear diophantine equations. For linear equations, the Chinese Remainder Theorem states more or less that distinct primes behave like random variables—they are completely independent of one another.

II. Next we studied quadratic equations. There are subtle relations between distinct primes: for instance, if p and q are primes of the form $4n + 1$, then $X^2 - pY - q$ has a solution in integers if and only if $X^2 - qY - p$ has a solution in integers. This is (part of) the famous law of quadratic reciprocity which was discovered by Gauss when he was 17 years old. Modern number theory has been greatly influenced by the search for higher reciprocity laws, i.e., involving cubic, quartic, … equations.

III. In the third section, we studied the p-adic numbers Q_p, which are the completion of Q with respect to the norm on Q defined by

$$|p^n a/b| = p^{-n},$$

for $a, b, n \in Z$, and $a \neq 0, b \neq 0$, $(a, b) = 1$. These numbers play an important role in diophantine equations. Namely, since $Q \subset Q_p$, a necessary condition for an equation to have any solutions in Q is that it have at least one solution in Q_p. It turns out to be pretty simple to check whether an equation can be solved in Q_p, so this observation is really useful. The Hasse Principle (also known as the local-global principle) states that for certain types of diophantine equations, for example equations of the form

$$a_1 X_1^2 + a_2 X_2^2 + a_3 X_3^2 = 0,$$

an equation has a solution in Q if and only if it has a solution in R and a solution in Q_p for all p.

IV. Finally, we gave a simple geometric proof of the Hasse principle in the above special case. We also showed examples of diophantine equations where the Hasse Principle fails; i.e., equations which have solutions in R and Q_p for all p, but have no solution in Q.

During the last three weeks of the program, the students began to devote about half their time to projects. Students chose between four projects and worked in groups of two to five people. One group drew pictures on Mathematica to demonstrate that the rings of integers in $Q(\sqrt{-d})$ are Euclidean when and only when $d = 1, 2, 5, 7, 11$. They also reported on Davenport and Chatland's (1950) work on determining the real quadratic Euclidean rings. The other projects were to demonstrate a special case of Fermat's Theorem, simulate the RSA cryptosystem on the computer, and report on Waring's problem (Niven and Zuckerman, 1991, Dickson, 1966, Narkiewicz, 1983). The group doing Waring's problem became fascinated with representation numbers of quadratic forms. One of the first things they noticed from computer data was that for $b = 1, 3,$ or 7, the number $b * 2^{2a+1}$ can be written in just one way as the sum of four squares. This turned out to be part of a larger pattern: if $r(k)$ denotes the number of ways k can be written as the sum of four squares, then $r(2k) = r(8k)$ for all k. These results, though available in the literature (Hardy and Wright, 1988, Chapter XX), are exciting to discover on one's own and give a very good sense of what mathematical research is like.

References

Chatland, H., and Davenport, H. (1950). Euclid's algorithm in real quadratic fields. *Canadian J. Math.* **2**:289–296.

Dickson, L. E. (1966). *History of the Theory of Numbers*. Chelsea Publication, New York, p. 717.

Hardy, G.H., and Wright, E.M. (1988). *An Introduction to the Theory of Numbers*. Clarendon Press, New York.

Narkiewicz, W. (1983). *Number Theory* World Scientific, Singapore.

Niven, I., and Zuckerman, H.S. (1991). *An Introduction to the Theory of Numbers*, 5th Ed. Wiley, New York.

A Seminar on Linear Optimization

LYNNE M. BUTLER

Haverford College

The goal of the seminar was to help every participant develop her mathematical talent and explore her mathematical interests. The seminar accommodated students with very different backgrounds and levels of preparation, sophistication, and even motivation. A menu of learning opportunities was offered from which each student selected what she needed, as articulated in a written personal contract that she negotiated (and renegotiated) with the seminar leader. The learning opportunities included: seminar-style interaction with the other participants, teaching assistant, and seminar leader; both routine and challenging problems to be done individually or collaboratively on background material as well as the seminar topic; four group projects planned in advance as well as two individual projects planned for participants with special needs. These learning opportunities are described in detail below. But first, the particulars of course organization and content are given.

Course organization

Our seminar met three times each week for discussion and project presentation (with both the seminar leader and teaching assistant in attendance) plus twice each week for collaborative work on problems (catalyzed by the teaching assistant). Each of the 13 participants selected her own level and nature of participation in a personal contract with the seminar leader. All but two seminar participants solved many of the 42 problems distributed among 15 assignments, and all but one participant contributed to one of four group projects. Individual projects were designed for two participants whose needs were not met by the group projects.

Course prerequisites and content

Seminar participants rediscovered how to apply multivariable calculus and linear algebra to solve a large class of optimization problems from game theory, operations research, and theoretical computer science. They did not need to know how to program, but they did have to develop their feel for geometry in n-dimensional Euclidean space. Thereby they gained a deep understanding of the simplex algorithm, which finds optimal solutions to linear programs.

During the first three weeks we focused on theory, from both algorithmic and geometric points of view, with an emphasis on the concept of duality. During this period we provided three assignments that reviewed linear algebra and multivariable calculus, as well as notes written by Harold Kuhn and eight assignments with problems on the following topics: Introduction to Linear Programming; Dual Programs in Canonical Form; The Pivot Operation; The Simplex Method; Geometry and Tableaux; The Pivot Operation in Standard Form; Bland's Anticycling Simplex Method; Duality for General Linear Programs.

In the last three weeks, student groups took over the seminar to present their independent work on four applications: finding optimal strategies for players in a zero sum game; finding least cost assignments; and finding a shortest path or maximum flow from source to sink in a network. Notes written by Harold Kuhn and seven assignments (three devised by students) were provided on the following topics: Matrix Games; The Assignment Problem; The Shortest Path Problem; Maximizing Flow in a Network.

An example of seminar-style interaction

Ideally, the leader of a seminar plans and directs activities that catalyze student discovery and exploration of the central ideas in an area of mathematics. One of the key definitions in the theory of linear programming is of an extreme point of a convex set. Proofs of the basic theorems of linear programming make use of the following formal definition: A point z in a convex set S is **extreme** if and only if the system

$$z = \lambda x + (1 - \lambda)y;$$

$0 < \lambda < 1; x, y \in S; x \neq y$ has no solution. That is, a point z in S is extreme if it is not interior to any line segment joining distinct points in S. During one seminar meeting, groups of participants formulated their own definitions of an extreme point and presented their findings to each other. All students saw many correct formulations of the same definition, and the ablest students devised examples that show the above definition is not equivalent to one that says a point z of S is extreme if there is a half space that intersects S only at z.

In our seminar, participants were first shown examples of extreme points of a closed, full-dimensional convex set in the plane (at corners of the set and on curved portions of its boundary) and examples of points in the set that are not extreme (in the interior and between corners on straight portions of its boundary). They were asked to break up into groups of three or four and formulate a definition of extreme that is consistent with the examples given.

The seminar might proceed as follows. The seminar leader and teaching assistant circulate. At first, a participant who is shy needs to be introduced to a group that is forming nearby. Later, members of a group that is being dominated by one individual are encouraged to express their ideas. When a group nears consensus, the seminar leader or teaching assistant raises questions or challenges the group to formulate a definition that covers convex sets in Euclidean space of any dimension. For example, a group that decides a point z of a convex subset S of the plane is extreme if there is a line that intersects S only at z is shown the example when S is a line segment in the plane and the line passes through the middle of S; after such a group has decided to use halfplanes instead of lines, they are encouraged to think about 3-dimensional space.

Each of the first few groups that settle on a definition is asked to nominate someone to write its definition on the blackboard. Each group that finishes early is challenged to think about whether the definitions that have been written on the board so far are equivalent. Meanwhile the seminar leader and teaching assistant track the progress of slower groups, helping them to see when their ideas lead to improved formulations of definitions already written on the blackboard. Near the end of the meeting, the seminar leader summarizes the findings by orchestrating a conversation about which definitions are equivalent, and she makes sure to acknowledge all the groups (not just the first) that helped formulate each of the two inequivalent definitions.

A sample challenging problem

Problems like the one below were worked on by individuals or groups; individuals and groups that floundered were given hints; imperfect solutions were shared for others to refine; solutions arrived at by one individual were checked by another; written solutions authored by one student were presented orally by another; participants were asked to devise similar problems, and to generalize the problem and their solution.

Problem

To increase security, the University of California at Berkeley has decided to station guards at some campus intersections. All potential locations are marked by a • in the campus map on page 61. A guard covers all of the road segments that meet at her intersection.

What is the smallest number of guards you can use to cover all the roads in the map on page 63? Where should you place them? Use König's Theorem (or something to which it is related) to prove that your answer is correct. What property of the map makes this proof viable?

Features of group projects

The group projects were designed to allow students to apply the theory of linear programming to important problems. Each group took over the seminar for one or two meetings to communicate its findings. Group members sometimes played the role of the seminar leader and teaching assistant. That is, they planned and directed activities that catalyzed student discovery and exploration of the ideas underlying the application studied by the group. Groups devised problems to be provided on assignments covering their topics. Group members often took turns explaining important concepts at the blackboard. Before doing so, they met several times with the seminar leader to plan and practice their presentations. Each was expected to try to express herself precisely, to answer probing questions, to supply examples and motivation throughout her presentation, and to think about ways to make her presentation more lively and interactive. Each member of the seminar was given a small gift after her turn leading the seminar (flowers, her favorite cookies, a math video or book or toy).

Groups formed over the course of the first three weeks, as participants discovered those with whom they collaborated easily. Topics were assigned to match the interests, talents, and preparation of the

Figure 1: Campus map

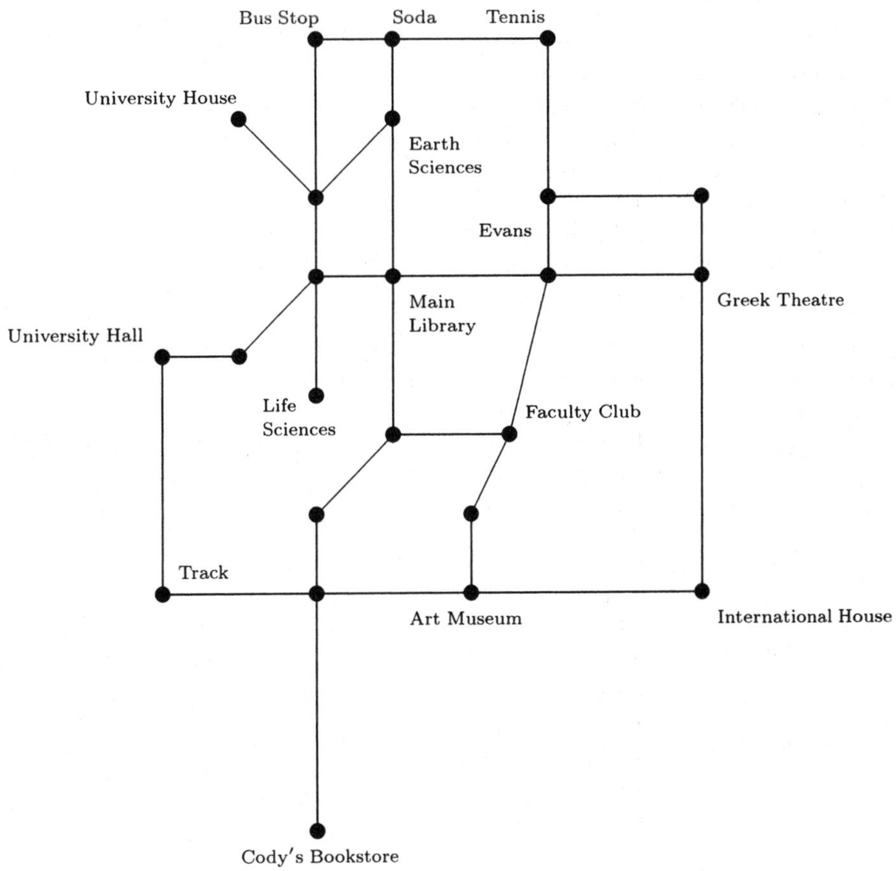

members. Shortest path and maximum flow were given to groups whose members were less experienced; matrix games and the assignment problem were given to groups that had some sophisticated members. Each group was provided with notes, problems, and references to stimulate its investigation. They each studied a journal article in its application area. These are described below:

Game Theory

NASH, J. (1951). Non-cooperative games. *Annals of Math.* **54**: 286–301.

Von Neumann's minimax theorem for zero sum games is most elegantly deduced from Farkas's lemma (the mathematics underlying duality theory of linear programming). On the third page of this paper, the existence of Nash equilibrium points in general sum games is deduced from Brouwer's fixed point theorem. This paper is the first, not the best written, reference. You might want to look at FRANKLIN, J. (1980). *Methods of Mathematical Economics*. Springer Verlag, New York.

Assignment Problem

KUHN, H.W. (1955). The Hungarian method for the assignment problem. *Naval Res. Logist. Quart.* **2**: 83–97.

This algorithm finds an optimal solution to an n person, n job assignment problem in $O(n^3)$ operations. To understand how it might have been discovered, formulate the assignment problem as a linear program, then study the dual. The Hungarian algorithm works with feasible solutions to the dual program and complementary solutions to the primal. When the solution to the primal becomes feasible, both the primal and the dual solutions are optimal.

Shortest Path

DIJKSTRA, E.W. (1959). A note on two problems in connection with graphs. *Numerische Mathematik* **1**: 269–271.

In one page of this two-page paper, Dijkstra describes his algorithm for finding the shortest path between two nodes in a graph. Try to imagine how he might have discovered this algorithm by first formulating the shortest path problem as a linear program and then studying the structure of the dual.

Maximum Flow

FORD, L.R., and FULKERSON, D.R. (1956). Maximal flow through a network. *Canadian Journal of Math.* **8**: 399–404.

This is a short paper on a famous algorithm to find a maximum flow in a network. Fulkerson told Kuhn that he was inspired by the Hungarian algorithm for the assignment problem, which works with feasible solutions to the dual program and complementary solutions to the primal. When the solution to the primal becomes feasible, both the primal and the dual solutions are optimal.

Anonymous student evaluations of the seminar

When asked whether the seminar was at an appropriate level, one participant responded:

> It required no advanced math, so everyone   was equally capable of understanding the material. However, the material was *not* at all presented in a simplistic manner (we did rigorous proofs that our algorithms worked).

When asked whether making a personal contract with the seminar leader was useful to them, some participants responded:

> Very! The contract gave me freedom to work in areas I needed and helped me stay on track. I plan to make contracts for all my classes this fall.

> The contract was helpful. I do not think I made mine very specific but it was helpful. The professor listened to my comments and tried to accommodate to what I wanted to do or not do. This (for some strange reason which I do not know) surprised me.

> No, my contract was not useful. I have a hard time identifying my strengths and weaknesses so I didn't know what I should concentrate on this summer.

When asked which activities in the seminar were effective, some participants responded:

> Lecture was extremely effective for me, because it encouraged me to think in class and to participate.

> Group work is always the most effective for me—I love working in a group, if teaching others or learning from others.

> The homework was time-consuming but insightful (almost always). The best was the solution set that was provided for each homework assignment.

> The project was the only thing which I was really able to get interested in.

When asked to comment on the seminar projects, some participants responded:

> I enjoyed my project, game theory. This topic interests me greatly. I liked how the last 2-3 weeks were basically dedicated to projects, so we could learn how the methods we learned in class applied to other topics.

> The group project forced me to develop a lecture that would be understandable to the rest of the class. I learned a lot about how to develop a topic in a logical manner, including as much motivation as possible for what is being done. Since I have thought of becoming a professor or teacher, this was a good chance for me to

see if I could deliver a substantial lecture, and also to see if I could enjoy lecturing. I found out I really do enjoy presenting material to others.

The seminar project I did was interesting. The way the project topics were given was also good. There seemed to be hard and not too hard projects so that everyone seemed to have found one that seemed appropriate for their level. We also got to pick our groups so we could work with people who were at the same level we were.

The project excited me, and I learned a lot about presenting, although at the time I was doubting my mathematical ability and wondered if I really knew anything at all about math!

What are Numbers?

Svetlana Katok

Pennsylvania State University

The goal of this seminar was to reveal the concept of *number* in modern mathematics. The title was inspired by Kirillov's book (1993). My original idea was to use it as the main source for the seminar, but unfortunately, it was too sophisticated for the students, and I ended up using only his general philosophy of consequent extensions of the notion of number, which appears in the beginning of this article. The material is presented in a series of problems with a skeletal framework that provides a context for them. In the seminar, students solved these problems and worked on projects, individually or in groups. Solutions and hints to some of these problems are included in the Appendix.

Introduction

We start with the following chain:

$$\mathbb{N} \subset \mathbb{Z} \subset \mathbb{Q} \subset \mathbb{R} \subset \mathbb{C} \subset \mathbb{H} \subset \mathbb{O}.$$

This chain of consequent extensions of the concept of number you probably know well enough, at least up to its fourth and fifth members. Symbols of this chain became standard notation for, respectively, sets of natural, integer, rational, real, complex numbers, quaternions, and Cayley numbers (also known as octaves). We discuss transitions from one term of this chain to the next, and show how the ideas underlying these transitions may lead to different, sometimes very unexpected, and beautiful theories.

The material is divided according to the following topics. Some of these topics are covered in the group projects, which are listed prior to the Appendix.

- Arithmetic: from natural numbers to integers, to rational numbers; arithmetic operations, groups, rings, fields.

- Analysis: from rational numbers to reals, a concept of completion, *p*-adic numbers.

- Rational numbers as an ordered field, a way of obtaining real numbers via cuts.

- More on real numbers, algebraic and transcendental numbers.

- Algebra: from real to complex numbers, algebraically closed fields; commutative, associative, and division algebras.

- The exceptional position of four algebras: real numbers, complex numbers, quaternions, and Cayley numbers—Hurwitz's Theorem.

The majority of the problems here are devoted to the theory of *p*-adic numbers, which are a remarkable diversion from the above mentioned chain. The material on which they are based can be found in the books by Borevich and Shafarevich (1996), Kirillov and Gvishiani (1982), and Koblitz (1977). The material on division and quaternion algebras can be found in Katok (1992). The book by Kantor and Solodovnikov (1989) was the primary source for the last topic.

Arithmetic: from $\mathbb{N}$ to $\mathbb{Z}$ and from $\mathbb{Z}$ to $\mathbb{Q}$; arithmetic operations, groups, rings, fields

Natural numbers can be added but not always subtracted; integers can be multiplied but not always divided. The urge to overcome these "inconveniences" leads to the transitions from $\mathbb{N}$ to $\mathbb{Z}$, and from $\mathbb{Z}$ to $\mathbb{Q}$.

Let us recall how to make those transitions. We want to subtract m from n. If $m \geq n$, then the answer is not in $\mathbb{N}$. We denote it by $n \ominus m$. We want all axioms of addition to hold in the extended set. Hence, we have to *identify* $n \ominus m$ with $(n+k) \ominus (m+k)$ for all $k \in \mathbb{N}$, and also with $(n-k) \ominus (m-k)$ for $1 \leq k < \min(m,n)$.

We see that the symbols $n_1 \ominus m_1$ and $n_2 \ominus m_2$ are identified if $n_1 + m_2 = n_2 + m_1$. Now let us consider all expressions

$$n \ominus m, m, n \in \mathbb{N}$$

with the given identification. We can add them by components

$$(n_1 \ominus m_1) + (n_2 \ominus m_2) = (n_1 + n_2) \ominus (m_1 + m_2),$$

and subtract them by the following rule:

$$(n_1 \ominus m_1) - (n_2 \ominus m_2) = (n_1 + m_2) \ominus (n_2 + m_1).$$

For instance,

$$(0 \ominus 0) - (m \ominus n) = n \ominus m.$$

1. Show that the equivalence classes of symbols

$$\{m \ominus n \mid m, n \in \mathbb{N}, \quad n_1 \ominus m_1 \sim n_2 \ominus m_2$$
$$\text{iff} \quad n_1 + m_2 = n_2 + m_1\}$$

form a group by addition, and that this group is isomorphic to $\mathbb{Z}$.

The procedure is completely analogous for the construction of the multiplicative group $\mathbb{Q}^*$ of non-zero rational numbers starting from a *semigroup* $\mathbb{Z} \setminus \{0\}$.

2. Prove that the equivalence classes of symbols

$$\{m : n \mid m, n \in \mathbb{Z} \setminus \{0\}, \quad m_1 : n_1 \sim m_2 : n_2$$
$$\text{iff} \quad m_1 n_2 = m_2 n_1\}$$

form a group by multiplication, and that this group is isomorphic to $\mathbb{Q}^*$.

In both of these exercises we extended the domain to obtain a group using the same principle. We introduced new symbols (negative numbers, fractions), and formed the *equivalence classes* in such a way that the laws which held in the original domain continued to hold in the extended domain. We shall see *equivalence relations* very often as we go alone.

3. Which of the following relations are equivalence relations?

a) Relation of equality of two numbers;

b) relation of similarity of two triangles;

c) relation of order on the real line;

d) relation of linear dependence in a vector space L of dimension $n > 1$;

e) relation of linear dependence on the set

$$L^* = L \setminus 0,$$

where L is a vector space.

Analysis: from $\mathbb{Q}$ to $\mathbb{R}$; a concept of completion, p-adic numbers

The real numbers are obtained from rationals by a procedure called *completion*. This procedure can be applied to any *metric space*, i.e., a space M with a distance function d on it. A sequence $\{r_n\} \in M$ is called a *Cauchy sequence* if for any $\epsilon > 0$ there exists $N > 0$ such that $n, m > N$ implies $d(r_n, r_m) < \epsilon$. If any Cauchy sequence in M has a limit in M, then M is called a *complete metric space*. If M is not complete, there exists a metric space $\overline{M}$ such that

1. $\overline{M}$ is complete;

2. $\overline{M}$ contains a subset $\overline{M}_0$ isometric to M;

3. $\overline{M}_0$ is dense in $\overline{M}$ (i.e., each point in $\overline{M}$ is a limit point for $\overline{M}_0$).

The elements of $\overline{M}$ are equivalence classes of Cauchy sequences in M: (two Cauchy sequences x_n and y_n are called equivalent if $d(x_n, y_n) \to 0$).

4. Prove that a metric space is complete if and only if the intersection of every descending sequence of closed balls whose radii approach zero consists of a single point.

Let us apply this construction to the rational numbers. We have the usual Euclidean distance between rational numbers:

$$(1) \qquad d(r_1, r_2) = |r_1 - r_2|.$$

The geometric interpretation of rational numbers as points on the "number axis" is obviously connected with this distance. It is easy to construct a Cauchy sequence of rational numbers which has no limit in $\mathbb{Q}$:

$$.1, .1011, .10110111, .1011011101111, \ldots$$

5. Prove that the rational numbers are represented by eventually periodic decimal fractions.

On the other hand, any point on the "number axis" can be represented by an infinite decimal fraction, and any Cauchy sequence of rational numbers has a limit that is an infinite decimal fraction. In other words, the construction of real numbers through infinite decimal fractions is equivalent to the completion procedure described above. We shall denote the set of real numbers by $\mathbb{R}$.

6. Prove that the following metric spaces are not complete, and construct their completions:

1. $\mathbb{R}$ with the distance

$$d(x, y) = |arctan\, x - arctan\, y|;$$

2. $\mathbb{R}$ with the distance $d(x, y) = |e^x - e^y|$.

7. On the set of closed intervals of the real line we define a distance by the formula:

$$d([a, b], [c, d]) = |a - c| + |b - d|.$$

Prove that the obtained metric space is not complete, and find its completion.

8. On the set $\{\Delta\}$ of closed intervals of the real line we define a distance by the formula:

$$d(\Delta_1, \Delta_2) = |\Delta_1| + |\Delta_2| - 2|\Delta_1 \cap \Delta_2|.$$

Prove that the obtained metric space is not complete, and find its completion.

9. Prove that the space of polynomials with real coefficients $\mathbb{R}[x]$ is not complete with respect to the following distances:

1. $d(P, Q) = \max_{[0,1]} | P(x) - Q(x) |$;

2. $d(P, Q) = \int_0^1 | P(x) - Q(x) | \, dx$;

3. $d(P, Q) = \sum_i |c_i|$, where

$$P(x) - Q(x) = \sum_i c_i x^i.$$

Notice that the Euclidean distance "came from" the Euclidean norm on $\mathbb{Q}$, which is the absolute value. Suppose we have a norm on a field F, i.e. a map denoted by $\| \ \|$ from F to the non-negative real numbers, such that

1. $\|x\| = 0$ iff $x = 0$,

2. $\|x \cdot y\| = \|x\| \cdot \|y\|$,

3. $\|x + y\| \leq \|x\| + \|y\|$.

Then we can define a distance $d(x, y) = \|x - y\|$. We say that this distance is *induced* by the norm $\| \ \|$. We say that the norm is *trivial* if $\|0\| = 0$ and $\|x\| = 1$ for all $x \neq 0$.

Now let us ask ourselves a question: is the Euclidean distance between rational numbers really the most "natural" one? Is there any other way to describe the "closeness" between them? It turns out that the answer to this question is YES!

Let us fix a prime number p. Then any rational number r can be uniquely written in the form

$$r = p^k \frac{m}{n},$$

where $k \in \mathbb{Z}$, and

$$(m, n) = (m, p) = (n, p) = 1.$$

This number k is denoted by $ord_p r$. If r is an integer, then $ord_p r$ is the greatest k such that $r \equiv 0 \pmod{p^k}$.

Definition.

$$\|r\|_p = \begin{array}{ll} p^{-ord_p r}, & \text{if } r \neq 0 \\ 0, & \text{if } r = 0. \end{array}$$

is called the *p-adic norm* of r.

10. Prove the following formulae:

1. $\|r_1 r_2\|_p = \|r_1\|_p \|r_2\|_p$;

2. $\|r_1 + r_2\|_p \leq \max(\|r_1\|_p, \|r_2\|_p)$;

3. if $\|r_1\|_p < \|r_2\|_p$ then $\|r_1 + r_2\|_p = \|r_2\|_p$.

We introduce a new (*p*-adic) distance on $\mathbb{Q}$ by the formula:

$$(2) \qquad d_p(r_1, r_2) = \|r_1 - r_2\|_p.$$

Definitions. A norm satisfying

$$\|r_1 + r_2\|_p \leq \max(\|r_1\|_p, \|r_2\|_p)$$

instead of the triangle inequality is called *non-Archimedian*. A norm which is not non-Archimedian is called *Archimedian*. A distance induced by a (non-) Archimedian norm is called (non-) Archimedian.

We sometimes let $\| \ \|_\infty$ denote the usual absolute value norm on $\mathbb{Q}$.

11. Prove that d_p is the distance function on $\mathbb{Q}$, i.e.

1. d_p is symmetric: $d_p(r_1, r_2) = d_p(r_2, r_1)$,

2. d_p is non-negative: $d_p(r_1, r_2) \geq 0$, and $d_p(r_1, r_2) = 0$ iff $r_1 = r_2$,

3. d_p satisfies the triangle inequality:

$$d_p(r_1, r_3) \leq d_p(r_1, r_2) + d_p(r_2, r_3).$$

It follows from Problem **10** that d_p satisfies a stronger inequality:

$$(3) \qquad d_p(r_1, r_3) \leq \max(d_p(r_1, r_2), d_p(r_2, r_3)).$$

Definition. A metric space with a distance satisfying (3) is called an *ultrametric* space.

Thus a field with a non-Archimedian norm is an ultrametric space.

12. Prove that all triangles in an ultrametric space are isosceles, and that the length of the base does not exceed the length of the side.

Let us define a disc of radius r (r is a non-negative real number) with center $a \in M$ (M is a metric space) :

$$D(a, r) = \{x \in M \mid d(x, a) \leq r\}.$$

13. Prove that if M is an ultrametric space, then any point in $D(a,r)$ is its center.

14. Prove that in any complete normed field, with a non-Archimedian norm, a series $\sum_n x_n$ converges if and only if $x_n \to 0$.

15. Prove that rational integers $\mathbb{Z}$ form a bounded set of diameter 1 with respect to the p-adic distance d_p.

Definitions. We say two metrics d_1 and d_2 are *equivalent* if a sequence is Cauchy with respect to d_1 iff it is Cauchy with respect to d_2. We say two norms are equivalent if they induce equivalent metrics. The symbol $\sim$ is used to represent the equivalence of norms.

16. Let $\| \ \|_1$ and $\| \ \|_2$ be two norms on a field F. Prove that $\| \ \|_1 \sim \| \ \|_2$ iff there exists a positive real number α such that $\|x\|_1 = \|x\|_2^\alpha$ for all $x \in F$.

17. Prove that if $0 < \rho < 1$, then the function on $\mathbb{Q}$

$$\|x\| = \begin{array}{ll} \rho^{ord_p x}, & \text{if } x \neq 0 \\ 0, & \text{if } x = 0, \end{array}$$

is a non-Archimedian norm. Notice that by Problem **16** it is equivalent to $\| \ \|_p$. (What is α?) What happens if $\rho = 1$, or if $\rho > 1$?

18. Prove that $\| \ \|_{p_1}$ is not equivalent to $\| \ \|_{p_2}$ if p_1 and p_2 are different primes.

19. Let $\| \ \| = | \ |^\alpha$, where α is a fixed positive number. Show that $\| \ \|$ is a norm iff $\alpha \leq 1$, and that in this case it is equivalent to $| \ |$.

20. Prove that two equivalent norms on a field are either both Archimedian or both non-Archimedian.

The field of p-adic numbers

We can apply our completion procedure to $\mathbb{Q}$ with respect to d_p to obtain a complete metric space denoted by $\mathbb{Q}_p$. Its elements are equivalence classes of Cauchy sequences with respect to the following relation: $\{a_i\} \sim \{b_i\}$ if $\|a_i - b_i\|_p \to 0$ as $i \to \infty$.

For any $x \in \mathbb{Q}$, let $\{x\}$ denote the constant Cauchy sequence. We have $\{x\} \sim \{y\}$ iff $x = y$. The equivalence class of $\{0\}$ is denoted by 0.

We define the norm $\| \ \|_p$ of an equivalence class a to be $\lim_{i \to \infty} \|a_i\|_p$, where $\{a_i\}$ is any representative of a.

21. Prove that if $\{a_i\}$ is a Cauchy sequence in $\mathbb{Q}$, then $\lim_{i \to \infty} \|a_i\|_p$ exists.

Remark. In going from $\mathbb{Q}$ to $\mathbb{R}$ the possible values of $\| \ \|_\infty = | \ |$ were enlarged to include all non-negative real numbers, but in going from $\mathbb{Q}$ to $\mathbb{Q}_p$ the possible values of $\| \ \|_p$ remain the same: $\{p^n\}_{n \in N \cup 0}$. This is the reason the p-adic norm is also called the *discrete valuation*.

Ostrowski Theorem. Every non-trivial norm $\| \ \|$ on $\mathbb{Q}$ is equivalent to $\| \ \|_p$ for some prime p or for $p = \infty$. □

Proof. We give a proof as a series of exercises.

I. Suppose there exists a positive integer n such that $\|n\| > 1$, and let n_0 be the least such n. Then we can write $\|n_0\| = n_0^\alpha$ for some positive real number α.

   a) Prove that for any $n \in \mathbb{Z}$, $\|n\| = n^\alpha$.

   b) Prove that for any $x \in \mathbb{Q}$, $\|x\| = x^\alpha$.

   c) Use Problem **19** to conclude that $\| \ \|$ is equivalent to the absolute value $| \ |$.

II. Suppose $\|n\| \leq 1$ for all positive integers n. Let n_0 be the least n such that $\|n\| < 1$.

   a) Why does such an n_0 exist?

   b) Show that n_0 must be a prime: $n_0 = p$.

   c) Prove that if q is a prime, $q \neq p$, then $\|q\| = 1$.

   d) Prove that for any positive integer a, $\|a\| = \rho^{ord_p a}$, where $\rho = \|p\| < 1$.

   e) Use Problem **17** to conclude that in this case $\| \ \|$ is equivalent to $\| \ \|_p$.

Given two equivalence classes a and b of Cauchy sequences, we define $a \cdot b$ to be the equivalence class of the sequence $\{a_i b_i\}$ where $\{a_i\} \in a$ and $\{b_i\} \in b$. The sum is defined similarly term-by-term.

22. Prove that the definition of the sum and the product of equivalence classes of Cauchy sequences does not depend on the choice of representatives.

The additive inverses are defined in an obvious way. For multiplicative inverses we need the following fact:

23. Prove that any Cauchy sequence is equivalent to one with no zero terms.

Then as a multiplicative inverse for $\{a_i\}$ we can take $\{1/a_i\}$ which is Cauchy unless $\|a_i\|_p \to 0$, i.e. unless $\{a_i\} = \{0\}$. Why?

Thus we obtain a *field* of p-adic numbers $\mathbb{Q}_p$.

Let us consider the following series

(1) $\qquad \dfrac{b_{-k}}{p^k} + \dfrac{b_{-k+1}}{p^{k-1}} + \ldots + b_0 + b_1 p + b_2 p^2 + \ldots.$

By the construction and a general theorem about completions, each series of the form (1) represents an element of $\mathbb{Q}_p$. The converse statement is also true.

Theorem. Every equivalence class a in $\mathbb{Q}_p$ has exactly one representative Cauchy sequence which is a sequence of partial sums of a series in the form (1). $\qquad \square$

Proof. First notice that it is sufficient to give a proof in the case $\|a\|_p \leq 1$.

24. Deduce the theorem from the statement for $a \in \mathbb{Q}_p$ with $\|a\|_p \leq 1$.

Now consider our Cauchy sequence $\{a_i\} \in a$, and let $N(j)$ be a natural number such that $\|a_i - a_{i'}\|_p \leq p^{-j}$ whenever $i, i' \geq N(j)$. We have for $i \geq N(1)$,

$$\begin{aligned} \|a_i\|_p &\leq \max(\|a_{i'}\|_p, \|a_i - a_{i'}\|_p) \\ &\leq \max(\|a_{i'}\|_p, 1/p), \end{aligned}$$

for all $i' \geq N(1)$. But $\|a_{i'}\|_p \to \|a\|_p$ as $i' \to \infty$. Hence for $i \geq N(1)$, we have $\|a_i\|_p \leq 1$.

25. If $x \in \mathbb{Q}$ and $\|x\|_p \leq 1$, then for any j there exists an integer n chosen from the set $\{0, 1, \ldots, p^j - 1\}$ such that $\|n - x\|_p \leq p^{-j}$.

Now apply Problem **25** to find a sequence c_j where $0 \leq c_j < p^j$ such that

$$\|c_j - a_{N(j)}\|_p \leq 1/p^j.$$

We want to show that $\{c_j\} \sim \{a_j\}$ and that it is a sequence of partial sums of a series of the form

(1). The first assertion follows from the estimate for $i \geq N(j)$,

$$\begin{aligned} \|c_i - a_i\|_p &= \|c_i - c_j + c_j - a_{N(j)} - (a_i - a_{N(j)})\|_p \\ &\leq \max(\|c_i - c_j\|_p, \|c_j - a_{N(j)}\|_p, \\ &\qquad \|a_i - a_{N(j)}\|_p) \\ &\leq \max(1/p^j, 1/p^j, 1/p^j) \\ &= 1/p^j. \end{aligned}$$

Hence $\|c_i - a_i\|_p \to 0$ as $i \to \infty$. The second assertion follows from the estimate

$$\begin{aligned} \|c_{j+1} - c_j\|_p &= \|c_{i+1} - a_{N(j+1)} + a_{N(j+1)} \\ &\quad - a_{N(j)} - (c_j - a_{N(j)})\|_p \\ &\leq \max(\|c_{i+1} - a_{N(j+1)}\|_p, \\ &\qquad \|a_{N(j+1)} - a_{N(j)}\|_p, \|c_j - a_{N(j)}\|_p) \\ &\leq \max(1/p^{j+1}, 1/p^j, 1/p^j) \\ &= 1/p^j. \end{aligned}$$

This proves that $c_j \equiv c_{j+1} \pmod{p^j}$, which is exactly equivalent to the claim.

26. Prove the uniqueness result: given $a \in \mathbb{Q}_p$ there is a unique Cauchy sequence $\{c_i\}$ for which

1. $0 \leq c_i < p^i \quad$ for $i = 1, 2, \ldots$.

2. $c_i \equiv c_{i+1} \pmod{p^i} \quad$ for $i = 1, 2, \ldots$.

Thus any element in $\mathbb{Q}_p$ can be represented by an infinite-to-the-left fraction in base p:

(2) $\qquad \ldots b_n \ldots b_2 b_1 b_0 . b_{-1} \ldots b_{-k}, \quad 0 \leq b_i \leq p-1.$

Arithmetic operations in $\mathbb{Q}_p$

Here are some examples in $\mathbb{Q}_7$:

$$\ldots 263 \times \ldots 154 = \ldots 455$$

$$\ldots 30.2 - \ldots 56.4 = \ldots 40.5$$

$$\ldots 421 : \ldots 153 = \ldots 615.$$

27. Prove that in $\mathbb{Q}_5$ there exists a square root of $-1 (= \ldots 44)$, and find its last three digits. How many such roots are there?

28. Solve the equation $x^2 - 6 = 0$ in $\mathbb{Q}_5$.

29. Solve the equation $x^2 - 7 = 0$ in $\mathbb{Q}_5$.

Let $\mathbb{Z}_p$ be the closure of $\mathbb{Z}$ in $\mathbb{Q}_p$, the set of *p-adic integers*.

30. Prove that $\mathbb{Z}_p = \{a \in \mathbb{Q}_p \mid \|a\|_p \leq 1\}$.

31. Prove that $\mathbb{Z}_p$ is the set of elements in $\mathbb{Q}_p$ of the form (2) in the previous section with $b_i = 0$ for $i < 0$.

The *p-adic* integers form a subring of $\mathbb{Q}_p$. Let $\mathbb{Z}_p^{\times}$ be the set of invertible elements in $\mathbb{Z}_p$ (also called *p-adic units*), i.e.

$$\mathbb{Z}_p^{\times} = \{x \in \mathbb{Z}_p \mid 1/x \in \mathbb{Z}_p\}.$$

Then $\mathbb{Z}_p^{\times} = \{x \in \mathbb{Z}_p \mid \|x\| = 1\}$.

32. Prove that the *p-adic* expansion of $a \in \mathbb{Q}_p$ has repeating digits onward from some point (i.e., it is eventually periodic) if and only if $a \in \mathbb{Q}$.

33. Prove that if $x \in \mathbb{Q}$ and $\|x\|_p \leq 1$ for every prime p, then $x \in \mathbb{Z}$.

The method used in the solution of Problem **28** is quite general.

Hensel's Lemma. Let $F(x)$,

$$F(x) = c_0 + c_1 x + \ldots + c_n x^n,$$

be a polynomial whose coefficients are *p-adic* integers. Let $F'(x)$,

$$F'(x) = c_1 + 2c_2 x + 3c_3 x^2 + \ldots + nc_n x^{n-1},$$

be the derivative of F. Let a_0 be a *p-adic* integer such that $F(a_0) \equiv 0 \pmod{p}$ and $F'(a_0) \not\equiv 0 \pmod{p}$. Then there exists a unique *p-adic* integer a such that

$$F(a) = 0 \quad \text{and} \quad a \equiv a_0 \pmod{p}.$$

(Note, in the special case of Problem **28**

$$\begin{aligned} F(x) &= x^2 - 6, \\ F'(x) &= 2x, a_0 = 1). \end{aligned}$$

$\square$

34. Explain why $\mathbb{Q}_{10}$ is not a field.

Additional problems on *p-adic* numbers

35. What is the cardinality of $\mathbb{Z}_p$? Prove your answer.

36. Let us consider a map

$$\varphi : \mathbb{Q}_p \to \mathbb{R},$$

which maps a *p-adic* number,

$$\ldots b_2 b_1 b_0 . b_{-1} b_{-2} \ldots b_{-k} \to b_{-k} \ldots b_{-2} b_{-1} . b_0 b_1 b_2 \ldots,$$

to a real number in base p. Prove that φ is a continuous map of $\mathbb{Q}_p$ onto $\mathbb{R}_+$, the set of non-negative real numbers, and that it maps $\mathbb{Z}_p$ onto the closed interval $[0, 1]$.

Notice that due to non-uniqueness of writing the real numbers in base p, this map is not 1-to-1.

37. Construct a 1-to-1 continuous map of $\mathbb{Z}_p$ onto a Cantor set such that the inverse map is also continuous.

38. Prove that for any finite p, any sequence of integers has a subsequence which is Cauchy with respect to $\| \ \|_p$.

Is it possible to determine by a *p-adic* expansion of a rational number whether it is positive or negative? The answer is YES, and it is given in the following problem.

39. Let $r \in \mathbb{Q}$. Prove that its *p-adic* expansion can be represented in the form $\ldots aaaaaab$, where the fragments a and b have the same number of digits. Prove that $r > 0$ is equivalent to $b > a$ in the usual sense (as integers written in base p).

40. Prove that it is impossible to introduce an *order relation* in $\mathbb{Q}$ such that

1. if $x > 0$ and $y > 0$, then $x + y > 0$;

2. if $x > 0$ and $y > 0$, then $xy > 0$;

3. if $x_n > 0$ and there exists a limit, $\lim_{n \to \infty} x_n = x$, then $x \geq 0$.

More on real numbers: algebraic and transcendental numbers

41. Prove that $\mathbb{Q}(^3\sqrt{2})$ is a field.

42. Prove that any finite extension of $\mathbb{R}$ is isomorphic to either $\mathbb{R}$ or $\mathbb{C}$.

43. Prove that $\mathbb{Q}(\sqrt{2})$ and $\mathbb{Q}(\sqrt{-2})$ are not isomorphic as fields.

Algebra: from reals to complex numbers, algebraically closed fields; Commutative, associative, and division algebras

Definition. Let F be a field of characteristic $\neq 2$, $a, b \in F^* = F \setminus \{0\}$, and

$$A = \left(\frac{a, b}{F}\right)$$

be a *quaternion algebra over F*, i.e., an algebra (a vector space and a ring) over F of dimension 4 with a basis $\{1, i, j, k\}$ where

$$i^2 = a, \quad j^2 = b, \quad k = ij = -ji.$$

Hamiltonian quaternions $\mathbb{H}$ correspond to the case $a = b = -1$.

44. Prove that multiplication of quaternions is associative.

45. Prove that if $a \in (F^*)^2$, then $A = \left(\frac{a, b}{F}\right) \approx M(2, F)$.

46. Prove that for any $\lambda \in F^*$, $\left(\frac{a, b}{F}\right) = \left(\frac{\lambda^2 a, b}{F}\right)$.

Definition. A quaternion algebra A is called a *division algebra* if each non-zero element of A has an inverse.

47. Prove that A is a division algebra if and only if $N(x) = 0$ only for $x = 0$.

48. Prove that if A,

$$A = \left(\frac{a, b}{F}\right),$$

is not isomorphic to $M(2, F)$, then A is a division algebra.

49. Prove that there are only two quaternion algebras over $\mathbb{R}$ up to an isomorphism:

$$A = \left(\frac{a, b}{\mathbb{R}}\right) \approx \mathbb{H},$$

if $a < 0, b < 0$, and otherwise

$$A = \left(\frac{a, b}{\mathbb{R}}\right) \approx M(2, \mathbb{R}).$$

50. Prove that $\overline{q_1 q_2} = \overline{q_2} \, \overline{q_1}$ and $\overline{q_1 + q_2} = \overline{q_1} + \overline{q_2}$.

Group Projects

- A non-Archimedian extension of the field of real numbers: Conway numbers and non-standard analysis

- Surreal numbers

- Representation of real numbers by continued fractions

- Alternative arithmetic on the numbers $a + bi$: double and dual numbers

- Transcendental numbers

- Quaternions and vector algebra in 3- dimensional real vector space

- From complex numbers to quaternions—the doubling procedure—Cayley numbers

- Proof of Hurwitz's theorem

- Quaternion algebras over $\mathbb{Q}$

- The connection with units in algebraic number fields

APPENDIX: Hints and Solutions to Selected Problems

6. After completion, $\left(-\frac{\pi}{2}, \frac{\pi}{2}\right)$ becomes $\left[-\frac{\pi}{2}, \frac{\pi}{2}\right]$, and $(0, \infty)$ becomes $[0, \infty)$.

7. Let $\Delta_n = [0, \frac{1}{n}]$. Then $\{\Delta_n\}$ is a Cauchy sequence, since given $\varepsilon > 0$, take $N > \frac{2}{\varepsilon}$, then for $m, n > N$,

$$d(\Delta_n, \Delta_m) < \varepsilon.$$

But $\{\Delta_n\}$ cannot converge to any closed interval of positive length. Hence the space is not complete, and we need to add "0-intervals," $a = [a, a]$, to complete it.

8. Any sequence of intervals with lengths converging to 0 is a Cauchy sequence; they all correspond to a single point α in the completion, such that $d(\alpha, \Delta) = |\Delta|$. To prove that this is the only point we have to add, show that for each sequence $\{\Delta_n\}$ for which $|\Delta_n|$ does not approach 0, there is a subsequence such that all intersections $\Delta_i \cap \Delta_j$ are not empty. Then use the fact that for intersecting intervals the distance coincides with the distance from Problem **7.**

9. The sequence $\{P_k\}$, where $P_k = \sum_{i=0}^{k}(\frac{x}{2})^i$, is a non-converging Cauchy sequence for all three distances.

16. Suppose $\| \ \|_1 \sim \| \ \|_2$, and $a \neq 0$ such that $\|a\|_2 \neq 1$ (we assume $\| \ \|$ is non-trivial), say $\|a\|_2 > 1$. Then $\|a\|_1 = \|a\|_2^\alpha$ for some α. Show that for all $x \in F$, $\|x\|_1 = \|x\|_2^\alpha$.

17. If $\rho = 1$ then you get a trivial norm. If $\rho > 1$, then you do not get a norm at all, since the triangle inequality will not hold.

18. The sequence $\{p_1^n\} \to 0$ in $\| \ \|_1$ but not in $\| \ \|_2$.

19. Prove the triangle inequality.

20. First prove that the norm is non-Archimedian iff $\|n\| \leq 1$ for any integer $n = 1 + \ldots + 1$, n times. Then use the sequence $\{1/n^i\}$.

21. If $a = 0$, then by definition $\lim_{i \to \infty} \|a_i\|_p = 0$. If $a \neq 0$, then there exists an $\varepsilon > 0$ such that for every $N > 0$, there exists an $i_N > N$ with $\|a_{i_N}\|_p > \varepsilon$. If we choose N large enough such that $\|a_i - a_j\|_p < \varepsilon$ for $i, j > N$, we have

$$\|a_i - a_{i_N}\| < \varepsilon \quad \text{for all } i > N.$$

Since $\|a_{i_N}\|_p > \varepsilon$, it follows from Problem **12** that the triangle with vertices $0, a_i, a_{i_N}$ is isosceles. Hence $\|a_i\|_p = \|a_{i_N}\|_p$. Thus, for all $i > N$, $\|a_i\|$ has the constant value $\|a_{i_N}\|$, which is then the $\lim_{i \to \infty} \|a_i\|$.

23. If $a_i = 0$, take $a_i' = p^i$. For a given $\varepsilon > 0$, choose $N > \max(M, \log_{1/p} \varepsilon)$, where M is chosen for a given sequence $\{a_n\}$. If $\{a_n\}$ is not equivalent to $\{0\}$, then we can always choose a subsequence which contains no zero terms.

24. Consider $a' = ap^k$, where $\|a\|_p = p^k$.

25. Let

$$x = \frac{a}{b},$$

written in lowest terms. Since $\|x\|_p \leq 1$, p does not divide b, $\{(b, p^j) = 1\}$. We can find integers s and t such that $sb + tp^j = 1$. Let $n = as$. Then

$$
\begin{aligned}
\|n - x\|_p &= \|as - (\frac{a}{b})\|_p \\
&= \|\frac{a}{b}\|_p \|sb - 1\|_p \\
&\leq \|sb - 1\|_p \\
&= \|tp^j\|_p \\
&= \|t\|_p / p^j \\
&\leq 1/p^j.
\end{aligned}
$$

We can add a multiple of p^j to n to get an integer between 0 and p^j for which $\|n - x\|_p \leq p^{-j}$ still holds.

28. Let $x = a_0 + a_1 \cdot 5 + a_2 \cdot 5^2 + \ldots$. Then

$$(a_0 + a_1 \cdot 5 + a_2 \cdot 5^2 + \ldots)^2 = 1 + 1 \cdot 5.$$

Comparing coefficients at 5^0, we get

$$a_0^2 \equiv 1 \pmod 5.$$

Hence $a_0 = 1$ or $a_0 = 4$. Let $a_0 = 1$. Comparing the coefficients at 5^1, we get

$$2a_1 \cdot 5 \equiv 1 \cdot 5 \pmod 5.$$

Hence $2a_1 \equiv 1 \pmod 5$, and $a_1 = 3$. Continuing this way, we determine all the a_i uniquely. We obtain $x = \ldots 4031$.

43. First show that any field isomorphism must act as an identity on $\mathbb{Q}$.

References

Borevich, Z. I., and Shafarevich, I. R. (1966). *Number Theory*. Academic Press, New York.

Kantor, I. L., and Solodovnikov, A. S. (1989). *Hypercomplex Numbers*. Springer-Verlag, New York.

Katok, S. (1992) *Fuchsian groups*. The University of Chicago Press, Chicago.

Kirillov, A. (1993). *Chto Takoe Chislo?*. Sovremennaia Matematika Dlia Studentov, Nauka, Moscow.

Kirillov, A., and Gvishiani, A. (1982) *Theorems and problems in functional analysis*. Springer-Verlag, New York.

Koblitz, N. (1977). *P-adic numbers, P-adic analysis, and Zeta Functions*. Graduate texts in mathematics **58**, Springer-Verlag, New York.

Algebraic Coding Theory

Vera Pless

University of Illinois

Introduction

I found teaching algebraic coding theory in the cooperative manner to be a rewarding and challenging experience. I arrived at the Mills Summer Mathematics Institute with two sets of notes: Basic material for codes, and Syndrome decoding, bounds and perfect codes. The notes contain basic definitions and theorems in coding; most of the theorems are without proofs. Each set of notes also includes problems and examples, which are designed to explicate the definitions and theorems.

The seminar was six weeks long, and it met three times a week for 90 minutes. For the first three weeks of the program, the 90-minute period was divided into two parts. In the first part, I lectured on the material in the notes. The rest of the time was spent working on the problems in the notes and on additional problems relevant to the material. The students divided themselves into three groups to work on the problems, and my teaching assistant and I circulated among them to discuss what they were doing, answer questions, and observe. This setup worked well. The students remained in these groups for the first three and a half weeks of the seminar. After that, each student chose a project from a list of four, and new groups were formed, according to their choice of project.

For reference, three copies of my text on coding (Pless, 1989), were left in the dorm. In addition, a list was provided of other texts that were available in the library. However, the students learned mainly by going over the notes and working out the problems. Occasionally, solutions to the problems were presented in the seminar by the students in order to illustrate an important point or to go further. Each group wrote up the solutions to all the problems, with individuals taking responsibility for different problems. Proving the theorems provided valuable exercises because it made the meaning of the theorems clearer to the students. I read all the homework solutions and wrote comments on them.

During the second week of the program, we introduced a third set of notes on an extension of the basic material introduced in the first two sets of notes. The topic was an important class of codes – cyclic codes. (Another possible extension would be the topic of weight distributions.) This material was more advanced and specialized than the material of the first two handouts. Learning about cyclic codes required learning a bit about groups, rings, and fields. Since the students had varied backgrounds in these subjects, they worked on filling in their backgrounds in a workshop organized by the teaching assistant.

After three weeks, I distributed the following original papers, which are in four distinct areas of coding. I also handed out two further sets of notes on combinatorial games and lexicodes as background material. The papers were:

Pless, V. (1986). Decoding the Golay Codes. *IEEE Trans. on Information Theory*, IT-32.
MacWilliams, F. J., and Sloane, N. J. A. (1973). On the Existence of a Projective Plane of order 10. *JCT(A)* **14**.
Brualdi, R. A., and Pless, V. (1993). Greedy Codes. *JCT(A)* **64**.
Sloane, N. J. A. (1981). Error-Correcting Codes and Cryptography. Reprinted from the *Mathematical Gardner*, edited by Klarner, Prindle, Weber, and Schmidt, Boston.

I spent one class period describing each of the four areas. Then students chose the paper that interested and, based on these choices, formed new work groups, two with two students each and two with three students each. Students devoted the next two weeks to understanding their papers, looking up new information in the library, dividing up the material among themselves, writing up reports, and rehearsing presentations. Although the laast weeks we concentrated on the projects, I also lectured on background material that was necessary for some of the projects and that I thought would be of interest to all.

In the last week of the seminar, each group turned in a final written report (I had read and commented on earlier drafts) and made a half-hour presentation to the rest of the students. The final presentations were very impressive. Generally the students used both the blackboard and overhead projector for their presentations. They split up their presentations in different ways, some with the same students alternating several times.

The first group could actually decode the large Golay code by hand. They explained the theoretical basis of this to the class and taught the members of the class how to decode. They clearly had fun. The second group understood and explained a quite complicated process of determining the number of vectors of weight 15 in a putative projective plane of order 10. The third group explained what impartial games were, what greedy codes were, and their relationship. The fourth group described both conventional cryptography and public key cryptography, gave a public key scheme using Goppa codes, and introduced authentication codes. All this represents a great deal of work, and their accomplishments showed how they had all really matured in this short time.

As I became more acquainted with the students, I learned about their hopes, future plans, and concerns. Some of these were about getting into graduate school, what happens in graduate school, the state of the job market, alternatives to an academic career, and combining a career and a family.

Below are the three sets of notes distributed to the students in the seminar. The problems are also included. In addition to the numbered exercises, the students also proved all of the theorems, with the exception of the Varshamov-Gilbert Bound.

Basic material for Codes

The practical problem underlying coding theory is the efficient and accurate transmission of information from one place to another. Practical uses at present include the following:

- The high-fidelity on compact disc recordings.

- Transmission of financial information.

- Data transfer from one computer to another or from memory to a central processor.

- Information transmission from a distant source such as weather or communications satellites; e.g., the Voyager spacecraft sent pictures of Jupiter and Saturn to Earth.

The physical medium through which information is transmitted is called a **channel**. Undesirable disturbances, called **noise**, may cause the received message to be different from the transmitted message. Noise may be caused by sunspots, lightening, folds in magnetic tape, meteor showers, competing telephone messages, random radio disturbance, poor hearing, poor speaking, etc.

Error-correcting codes deal with the problem of detecting and correcting transmission errors caused by noise on a channel. Improving the channel is an engineering problem which we have no control over. We assume our channel is quite good. We will be mainly concerned with binary codes so we are in the situation where the information is transmitted by a sequence of zeros and ones. We call a 0 or a 1 a **digit**. A **word** is a sequence of digits. The **length** of a word is the number of digits in the word. A word is transmitted by sending its digits, one after the other, across a binary channel. A binary code C is a set of words. A **block code** is a set of words of a fixed length n. Words that belong to a code will be called **codewords**. The standard way for combating errors is to add redundancy. We do this in coding also, so a codeword can be regarded as consisting of information digits and redundancy digits. Many businesses commonly add check digits to identification numbers; these are extra digits which are used to check the correctness of data or account numbers. **Encoding** is the process of adding redundancy to the information digits. **Decoding** is the process of finding the "most likely" codeword sent from the word received. We want the encoding to be easy and the decoding to be fast. We also want to add the least amount of redundancy to correct a fixed number of errors. We assume that errors are randomly distributed. This is not always the case in practice; often errors occur in bursts. However, it is best to understand how codes correct random errors first and then to consider those modifications needed to correct bursts of errors.

If u and v are binary words of the same length, then $u + v$ is the binary word whose digits are the sums of the digits of u and v (mod 2). Thus, if $u = (0, 1, 0, 0, 1, 1)$ and $v = (1, 0, 0, 1, 1, 1)$, then $u + v = (1, 1, 0, 1, 0, 0)$. Even though any set of words is a code, more is known about linear codes and the block codes which are used are linear codes, so we will be concerned with only linear codes. A binary code C is **linear** if $u + v$ is in C whenever both u and v are in C. This means that the zero vector 0 is in C as $u + u = 0$ for any word u. If all the codewords in C have length n, linearity is the same as saying that C is a vector subspace of the vector space of all binary n-tuples. Any vector subspace has a certain dimension k, and this is precisely the length of the information portion of a codeword.

As an (n, k) code C is a k-dimensional subspace, it has a basis consisting of k linearly independent words. The $k \times n$ matrix whose rows are these words (in any order) is called a generator matrix for C. Clearly C has many generator matrices. For example, a $(6, 3)$ code C can be described by the generator matrix G given below,

$$G = \begin{pmatrix} 1 & 0 & 0 & 0 & 1 & 1 \\ 0 & 1 & 0 & 1 & 0 & 1 \\ 0 & 0 & 1 & 1 & 1 & 1 \end{pmatrix}. \qquad (1)$$

The first three positions of C are information positions, and the last three positions are the redundancy positions. We can encode any of the 2^3 messages consisting of 3 digits by adding the appropriate rows of G. Thus 0,1,1 is encoded as the codeword $(0, 1, 1, 0, 1, 0)$. Whenever a generator matrix is of the form (I, A), where I is a $k \times k$ identity matrix, as G is above, then it is said to be in **standard form**. We see that it is easy to encode a code which has a generator matrix in standard form. If C is an (n, k) code and we permute the n coordinate places, then we obtain another code C'. Any code C' obtained from C by a coordinate permutation is said to be **equivalent** to C.

Theorem 1. Any code is equivalent to a code which has a generator matrix in standard form. □

The other common way of describing a code is by parity check equations or its parity check matrix. The former, for the example above, is the set of equations:

$$\begin{aligned} a_4 &= a_2 + a_3, \\ a_5 &= a_1 + a_3, \text{ and} \\ a_6 &= a_1 + a_2 + a_3. \end{aligned}$$

This means that a vector $(a_1, a_2, a_3, a_4, a_5, a_6)$ is in C if and only if the a_i satisfy these parity check equations, where a_1, a_2, and a_3 can be specified arbitrarily. This is the same thing as being orthogonal to the following parity check matrix H,

$$H = \begin{pmatrix} 0 & 1 & 1 & 1 & 0 & 0 \\ 1 & 0 & 1 & 0 & 1 & 0 \\ 1 & 1 & 1 & 0 & 0 & 1 \end{pmatrix}. \qquad (2)$$

The condition that a vector $(a_1, a_2, a_3, a_4, a_5, a_6)$ is orthogonal to the three rows of H is seen to be the same as the parity check equations above. From this view, C is specified as being the complete set of vectors which are orthogonal to all the vectors in the code whose generator matrix is H. This code is denoted, as usual, by $C^{\perp}$, and is called the **dual code** of C. A generator matrix of $C^{\perp}$ is called a

parity check matrix of C. In other words, if C is an (n, k) code, then $C^{\perp}$ consists of all vectors orthogonal to the vectors in C with respect to the usual inner product, and $C^{\perp}$ is an $(n, n - k)$ code. If $C = C^{\perp}$, then C is said to be **self-dual**, and n must be even as $k = n/2$. If a generator matrix of C has the form (I, A), then a parity check matrix of C is given by (A^t, I). This can be verified by direct computation, and an example of this is given by G and H above.

The process of encoding is a procedure for adding redundancy to the message. The generator and parity check matrices provide two ways of doing this. Using the generator matrix, the linear combination of the basis vectors giving the message also gives the redundancy. Using the parity check matrix, the redundancy digits are solved in terms of the message or information digits. There are other methods of encoding for specific families of codes like cyclic codes, but in general the problem of encoding is an easy one. Not so the problem of decoding, which is the process of retrieving the message if some digits have been changed in the codeword during transmission over a noisy channel. This is a difficult and central problem in coding and has led to many ingenious decoding schemes for individual codes, for specific families of codes, or for all error-correcting codes. It has also motivated the development of the theory of error-correcting codes as such knowledge is useful in decoding.

The **weight** of a codeword x, denoted by $w(x)$, is the number of non-zero components it has. The **distance** between two codewords x and y, denoted by $d(x, y)$, is the weight of $(x + y)$. The **minimum weight** of a code, denoted by d, is the weight of the non-zero codeword of smallest weight. For a linear code, d is also the minimum non-zero distance between codewords. The terminology (n, k, d) stands for an (n, k) code of minimum weight d. The minimum weight is a very important concept as it can be shown that a code with minimum weight d can correct $\left[\frac{d-1}{2} \right]$ errors. What is actually true is the following.

Theorem 2. If C is an (n, k, d) code, then any word of weight $t = \left[\frac{d-1}{2} \right]$, or less, is "closest" to a unique codeword. This is considered the most likely word sent. □

So, given n and k, we want d as large as possible, or given n and d, we want k as large as possible. These are very hard to do, and one of the main problems in coding has been the construction of such codes.

The following is another useful theorem.

Theorem 3. If C is an (n, k, d) code with parity check matrix H, then any $d - 1$ or fewer columns of H are linearly independent and some d columns of H are dependent.   □

Exercises

1. Let C be a code with generator matrix

$$G = \begin{pmatrix} 1 & 0 & 0 & 1 \\ 0 & 1 & 1 & 1 \end{pmatrix}.$$

a) If $v_1 = (0, 1, 0, 0)$, $v_2 = (1, 0, 1, 0)$, and $v_3 = (0, 1, 1, 0)$ are three received words, find a most likely sent codeword.

b) For these three received words, is there exactly one most likely codeword?

c) Find a parity check matrix for C.

2. a) Find the minimum weight of the code C with generator matrix (1).

b) If $v_1 = (1, 0, 1, 0, 0, 0)$, $v_2 = (0, 0, 1, 0, 0, 1)$, and $v_3 = (0, 0, 0, 0, 1, 1)$ are three received words, find a most likely sent codeword.

c) For these three received words, is there exactly one most likely codeword?

3. Show that any two (7,4,3) codes are equivalent.

4. Consider the following:

Let C be the set of 7-tuples with the ith digit in the ith labeled region so that the number of ones in any circle is even. For example $(1,1,0,0,0,1,1)$ is in C as we can see from the picture.

a) Show that C is a linear code.

b) How many codewords are in C?

c) Construct a generator matrix in standard form for C.

d) Construct a parity check matrix for C.

e) Decode the following received words:

$$\begin{aligned} v_1 &= (0, 0, 1, 1, 1, 0, 1), \\ v_2 &= (1, 1, 1, 1, 1, 0, 0), \\ v_3 &= (1, 0, 0, 0, 0, 1, 0). \end{aligned}$$

f) Show that any received word is either in C or is distance 1 to a unique codeword in C. Deduce from this that C can correct all single errors.

g) Can C correct any double errors?

h) Give a decoding algorithm for C.

i) Is it possible to label the portions of the circles so that the code C' given by the rule above does not have a generator matrix in standard form?

Syndrome decoding, bounds, and perfect codes

In order to decode a specific code, theoretically one can list all the codewords and compare the received word with each codeword to find the closest one. This is impractical for all but the smallest codes, even on a computer. A widely used code is the Golay (23,12,7) code, and I would not care to compare a 23-length binary vector with all (or even half of the) 2^{12} vectors in this code. So this is not the procedure used. There are clever decoding algorithms for specific codes like the Golay code, or for families of codes. These are better (for the codes for which they are designed) than syndrome decoding, which is a decoding procedure for all codes. However, syndrome decoding (in contrast to table look-up) is good enough to be used at times, so we describe it. First we need to know about cosets.

If C is a code and a is any word, then $\{a + c \mid c \text{ in } C\}$ is a **coset** of C. If a is in C, this coset is C itself.

Theorem 4. An (n, k) code has 2^{n-k} cosets. Each coset has 2^k words. Any two cosets are either equal or disjoint. The cosets of C partition the set of 2^n words.   □

The word of smallest weight in a coset (not the code) is called a **coset leader**. The **weight** of a **coset** equals the weight of any coset leader. If H is an $(n-k) \times n$ parity check matrix of an (n,k) code C, and u is a length n word, then the column vector formed by taking the usual inner product of u with the rows of H is called the **syndrome** of u. For example, if H is as in (2), then the syndrome of $(0,1,1,1,0,0)$ is

$$\begin{pmatrix} 1 \\ 1 \\ 0 \end{pmatrix}.$$

Theorem 5. All words in the same coset have the same syndrome. Words in different cosets have different syndromes. A word has syndrome equal to the zero vector iff it is in the code. All 2^{n-k} syndromes appear, one for each coset. $\square$

The weight of the coset of largest weight is called the **covering radius** of the code.

We begin syndrome decoding of a code C by finding a parity check matrix H for C. Then we generate a list of coset leaders and their syndromes. If u is a received word, we find its syndrome by taking inner products with the rows of H. We then look up this syndrome in our list and let e be the coset leader corresponding to this syndrome. Then we decode u to $u + e$, which must be in our code.

Theorem 6. Syndrome decoding is maximum likelihood decoding. This means that each word is decoded to the unique codeword closest to it, if there is one. If not, it is decoded to some codeword that is as close to it as any other codeword. $\square$

By turning this procedure around, we can devise a clever decoding algorithm for certain single error-correcting codes. For example, write the numbers 1 through 7 in binary columns of a parity check matrix H,

$$H = \begin{pmatrix} 0 & 0 & 0 & 1 & 1 & 1 & 1 \\ 0 & 1 & 1 & 0 & 0 & 1 & 1 \\ 1 & 0 & 1 & 0 & 1 & 0 & 1 \end{pmatrix}. \qquad (3)$$

Note that the syndrome of any vector is the position in error in binary. This code is called a Hamming $(7,4,3)$ single error-correcting code. Similar things can be done for other lengths. In fact, there is a family of $(2^r - 1, 2^r - 1 - r, 3)$ codes called Hamming single error-correcting codes with analogous nice decoding.

In order to prepare for the syndrome decoding of some code C, one needs a list of syndromes and corresponding coset leaders. One way of doing this is to write down all the words in each coset, and find a coset leader and syndrome. We did not do that for the Hamming codes, and sometimes one can get the information needed for syndrome decoding of other codes without such a big effort.

It is quite often interesting to determine how many codewords can be in a code of a given length n and minimum distance d. First we need a few more facts about the distance between two words x and y. Recall that $d(x, y) = wt(x + y)$. The distance function satisfies the following (and so is a metric):

1) $d(x, x) = 0$ for all x.
2) $d(x, y) = d(y, x)$ for all x, y.
3) $d(x, y) \le d(x, z) + d(z, y)$ for all x, y, z.

Only the last requires any proof. Note that if C is a code of minimum distance $d = 2t + 1$, then no word is distance t from any two codewords.

The following is a more formal version of Theorem 2. It follows from the fact that the number of words at distance t or less from any word u is

$$\binom{n}{0} + \binom{n}{1} + \cdots + \binom{n}{t}.$$

Theorem 7. (The Hamming Bound). If S is a code of length n and minimum weight $d = 2t + 1$ or $d = 2t + 2$, i.e. $t = \lfloor \frac{d-1}{2} \rfloor$, then

$$|C| \left(\binom{n}{0} + \binom{n}{1} + \cdots + \binom{n}{t} \right) \le 2^n,$$

which means that

$$|C| \le \frac{2^n}{\left(\binom{n}{0} + \binom{n}{1} + \cdots + \binom{n}{t} \right)}.$$
$\square$

Theorem 8. (The Singleton Bound). If C is an (n, k, d) code, then $d - 1 \le n - k$. $\square$

How large a dimension can a code with $n = 15$ and $d = 5$ have? By the Hamming bound,

$$\begin{aligned} |C| &\le \frac{2^{15}}{\binom{15}{0} + \binom{15}{1} + \binom{15}{2}} \\ &= \frac{2^{15}}{121} < \frac{2^{15}}{64} = \frac{2^{15}}{2^6} = 2^9, \end{aligned}$$

so $k \le 8$. We do not know whether such a code exists and, if it does, how to construct it. We try first to construct a (15,6,5) code by its parity check matrix H. We can suppose that $H = (I_9, A)$, where A is an 9×6 matrix such that every 4 columns of H are linearly independent. Hence the columns of A must have weight 3 or more. We can certainly choose the first three columns of A to be

$$\begin{pmatrix} 1 \\ 1 \\ 1 \\ 0 \\ 0 \\ 0 \\ 0 \\ 0 \\ 0 \end{pmatrix} \begin{pmatrix} 0 \\ 0 \\ 0 \\ 1 \\ 1 \\ 1 \\ 0 \\ 0 \\ 0 \end{pmatrix} \begin{pmatrix} 0 \\ 0 \\ 0 \\ 0 \\ 0 \\ 0 \\ 1 \\ 1 \\ 1 \end{pmatrix}.$$

But can we go further? Well, we want to choose the next column so that it is not equal to any of the first 12 columns of the H we have already chosen or to combinations of them two or three at a time. This will be possible if

$$\binom{12}{1} + \binom{12}{2} + \binom{12}{3} \le 2^9 - 1,$$

which is the number of non-zero length 9 vectors. As this is so, we can proceed.

The next theorem gives a lower bound on the maximum dimension of a code with a fixed length n and minimum distance d.

Theorem 9.   (The Varshamov-Gilbert bound). There exists an (n, k, d) code if

$$\binom{n-1}{0} + \binom{n-1}{1} + \ldots + \binom{n-1}{d-2} < 2^{n-k}$$

or

$$\frac{2^{n-1}}{\binom{n-1}{0} + \binom{n-1}{1} + \cdots + \binom{n-1}{d-2}} \le |C|.$$

$\square$

When d is odd and equality holds in the Hamming bound, then every vector in the space of binary n-tuples is within distance $t = \left\lceil \frac{d-1}{2} \right\rceil$ or less to a unique vector in the code. A code with this property is called **perfect**. A perfect code can correct all t or fewer errors and no other errors. The Hamming (7,4,3) code is perfect (Why?). Since perfect codes are very good, people spent much time looking for them. The first thing to do is to find n and t so that the Hamming bound is satisfied. The smallest value of n where a perfect multiple ($t > 1$) error-correcting code can occur is $n = 23, k = 12, d =$

7. Inspired by these parameters Golay constructed his famous (23,12,7) code. Even though there are other parameters which satisfy the Hamming bound, it was shown (and this was not easy to do) that the Golay parameters are the only ones possible for a perfect multiple error correcting code to exist. Further, any two codes with these parameters are equivalent. Since the Golay codes are so important, and have been used in several practical settings, we will give a generator matrix for this code and show, without using a computer, that the minimum weight is 7.

It is actually much easier to show that the **extended** (add an over-all parity check to each vector) Golay (24,12) code has minimum weight 8. Consider the generator matrix $G = (I, A)$ in the figure.

Let C be the (24,12) code with generator matrix G. We will show that C has minimum weight 8. Note that C is self-dual. Further, it is easy to show that since C is self-dual and the weights of the basis vectors in G are all divisible by 4, all weights in C are divisible by 4. Hence we have to show only that C cannot contain any codewords of weight 4. Since $A = A^\perp$ and C is self-dual, another generator matrix of C is (A, I). If c is any codeword, we let the left weight of c denote the weight of its first 12 digits and right weight of c denote the weight of its last 12 digits. By inspecting the rows of G we see that there is no codeword whose left weight is with right weight 3. A simple computation shows that there is no codeword with left weight 2 and right weight 2. Using the generator matrix (A, I), we see that there is no codeword with left weight 3 and right weight 1. So C is a (24,12,8) code, and, if we **puncture** (delete) any coordinate position, we get a (23,12,7) code.

Exercises

5. Decode the words in Exercises 1a) and 2b) by syndrome decoding.

6. Give the parity check matrix for a (15,11,3) single error-correcting code, where the syndrome of the received word is the position in error.

7. If H is a parity check matrix for a code with covering radius r, show that r is the smallest number so that any non-zero syndrome is a linear combination of r or fewer columns of H.

8. Find the covering radius of the following codes:

∞	0	1	2	3	4	5	6	7	8	9	10	∞	0	1	2	3	4	5	6	7	8	9	10
1												1	1	1	1	1	1	1	1	1	1	1	1
	1											1	1	1		1	1	1				1	
		1										1	1		1	1	1				1		1
			1									1		1	1	1			1		1	1	
				1								1	1	1				1		1	1		1
					1							1	1			1		1	1		1	1	1
						1						1			1		1	1		1	1	1	
							1					1		1		1	1		1	1	1		
								1				1	1		1	1		1	1	1			
									1			1	1		1	1		1	1	1			
										1		1	1		1	1		1	1	1			1

a) The (6,3,3) code in our example.

b) The (7,4,3) Hamming code.

c) Any Hamming code.

9. Does there exist a (12,7,5) binary code?

10. Show that a binary (15,6,5) code exists by constructing its parity check matrix.

11. Find a generator matrix for the code with parity check matrix (3).

12. Give upper and lower bounds on the largest dimension of a code with $n = 9$ and $d = 5$.

The group of a code and cyclic codes

If we consider all the permutations on n objects (which form a group called S_n) as acting on n coordinates, some send an (n, k) code C onto itself and some send C onto an equivalent code.

The set of permutations which send C onto itself is a group. Why? If C is a binary code, this group is called the **group** of C. Knowledge about the group of a code is often very useful in many different ways. If we know C, by either a generator matrix or parity check matrix, then it may not be easy to find the entire group of C. Often it is easy to find a subgroup of the group of C.

The group of a code sends all vectors of a fixed weight in the code onto themselves. Why? This group is clearly a subgroup of S_n.

It is customary (for good reasons) to consider only cyclic codes of odd length n, and for such, we label the n coordinates as $0, 1, \ldots, n - 1$. Then C is a **cyclic code** if the group of C contains the cyclic group generated by the coordinate permutation $i \to i + 1 \ (mod \ n)$. If C is a cyclic code and $c = (c_0, \ldots, c_{n-1})$ is in C, then so are all cyclic shifts of c, i.e. $(c_{n-1}, c_0, \ldots, c_{n-2}), \ldots$.

As we will see, it is useful to identify codewords in cyclic codes with polynomials. The coordinates represent the powers of x, and the component in a coordinate position gives the coefficient of the power of x. That is why we labeled the coordinates 0,1,...,n-1. So the vector

$$\begin{array}{ccccccc} 0 & 1 & 2 & 3 & 4 & 5 & 6 \\ (\ 0 & 1 & 1 & 0 & 1 & 0 & 0 \ ) \end{array}$$

corresponds to the polynomial $x + x^2 + x^4$. The reason for setting up this correspondence is that we will find polynomial multiplication useful. If our cyclic code is of length 7, then codewords correspond to polynomials of degree at most 6. But if we multiply two polynomials of degree at most 6, we might get a polynomial of degree higher than 6, which would not correspond to any vector of length 7. To overcome this, we multiply polynomials of degree at most 6 $mod(x^7 - 1)$. For example,

$$(1 + x^2 + x^4)(x + x^3 + x^4) = (1 + x^4 + x^5).$$

Let F be the field $GF(2)$; then $F[x]$ is the set of all binary polynomials. This is a commutative ring with unit where we can add and multiply polynomials. Also the division algorithm holds in $F[x]$; that is, if $a(x)$ and $b(x) \neq 0$ are polynomials, then there are unique polynomials $s(x)$ and $r(x)$ so that

$$a(x) = b(x)s(x) + r(x),$$

where the degree of $r(x)$ is less than the degree of $b(x)$.

Let $\mathbf{R_n} = F[x]/(x^n - 1)$. Then $\mathbf{R_n}$ is also a commutative ring with unit. Further, every ideal I in $\mathbf{R_n}$ consists of the multiples of some polynomial, and that polynomial is said to generate I. A ring with this property is called a **principal ideal ring** (PIR).

What makes this set-up so useful is that cyclic codes of length n and the ideals in $\mathbf{R_n}$ are the same under the correspondence we set up above. We look at the vectors corresponding to the polynomials below to see why this is so.

	vectors	polynomials
$c =$	$(0\ 1\ 1\ 0\ 1\ 0\ 0)$	$x + x^2 + x^4$
	$(0\ 0\ 1\ 1\ 0\ 1\ 0)$	$x^2 + x^3 + x^5$
	$(0\ 0\ 0\ 1\ 1\ 0\ 1)$	$x^3 + x^4 + x^6$

So we identify cyclic codes of length n with ideals in $\mathbf{R_n}$. If $\mathcal{C}$ is generated by a polynomial $f(x)$, we say $\mathcal{C} = <f(x)>$.

Theorem 10. If $\mathcal{C}$ is an ideal in $\mathbf{R_n}$, the unique monic polynomial $g(x)$ of smallest degree in $\mathcal{C}$ generates $\mathcal{C}$. $\square$

Further, $g(x)$ divides $(x^n - 1)$ and, conversely, if a polynomial in $\mathcal{C}$ divides $x^n - 1$, then $g(x)$ is the polynomial of lowest degree in $\mathcal{C}$.

This polynomial is called the **generator polynomial** of $\mathcal{C}$.

Theorem 11. If $\mathcal{C}$ has a generator polynomial $g(x)$ of degree $n - k$, then the dimension of $\mathcal{C}$ is k. $\square$

If
$$g(x) = g_0 + g_1 x + g_2 x^2 + \ldots + g_k x^k,$$

then the following matrix is a generator matrix of $\mathcal{C}$.

$$\begin{pmatrix} g_0 & g_1 & g_2 & \cdots & g_{n-k} & 0 & \cdots & 0 \\ 0 & g_0 & g_1 & \cdots & & g_{n-k} & 0 \cdots & 0 \\ \vdots & & & & \vdots & & & \vdots \\ 0 & & \cdots & 0 & g_0 & g_1 & \cdots & g_{n-k} \end{pmatrix}$$

This means that to find all binary cyclic codes of length n, we only need to factor $x^n - 1$ over GF(2).

A polynomial is called **reducible** if it can be factored. Otherwise, it is called **irreducible**. We really want to factor $x^n - 1$ into irreducible factors.

Cyclic codes can be implemented on an easily obtainable device called a shift register. This makes the encoding of a cyclic code more efficient than the encoding of an ordinary linear code and also helps with the decoding of these codes.

Exercises

13. a) Why is the set of coordinate permutations sending a code on to itself a group?

b) Are the permutations in the group of $\mathcal{C}$ linear transformations on $\mathcal{C}$?

c) Give an example of a linear transformation on a code which is not a coordinate permutation.

14. Find the order of the group of the code with the following generator matrix.

a)
$$G = \begin{pmatrix} 1 & 1 & 0 & 0 \\ 0 & 1 & 1 & 0 \\ 0 & 0 & 1 & 1 \end{pmatrix}$$

b)
$$G = \begin{pmatrix} 1 & 1 & 0 & 0 \\ 0 & 0 & 1 & 1 \end{pmatrix}$$

Are these groups transitive?

15. Why are cyclic codes and ideals in R_n the same?

16. Why is R_n a PIR?

17. Find all binary cyclic codes of length 3; that is, give their dimensions and generating polynomials.

18. Do the same thing for all cyclic codes of length 5.

19. What is the dimension of the cyclic code generated by the vector $(0, 1, 1, 0, 1, 0, 0)$?

References

Pless, V. (1989). *Introduction to the Theory of Error Correcting Codes*. Wiley, New York.

Quadratic Reciprocity and Continued Fractions

LYNNE WALLING

University of Colorado, Boulder

The seminar met for 90 minutes three times a week, for six weeks. The first four meetings were lectures on some basic definitions and results with particular attention to representation of numbers as positive definite quadratic forms. For the remainder of the seminar, the students worked in small groups of two to four solving exercises, reading research papers, and making presentations.

Exercises

The students formed three teams, and each team solved every third problem. That is, team k solved all problems congruent to k modulo 3. The first set of exercises were from Niven, Zuckerman, and Montgomery's book, *An Introduction to the Theory of Numbers*. These exercises involved equivalence and reduction of binary quadratic forms. Since the students were initially nervous about making presentations at the chalkboard, they each submitted written solutions, which were marked and returned.

The other sets of exercises appear below. They involved determining which primes can be written as a sum of two squares, showing that every positive integer is a sum of four squares, and using continued fractions to find solutions to Pell's equation. The students rotated in presenting their team's solutions. Also, each student was required to organize the results on the sum of two squares into lemmas, propositions, theorems, and proofs. These papers were marked, and each student was asked to rewrite a particular portion of her paper in a format acceptable for publication.

Research Papers

A list of research papers, which I thought were reasonably accessible to undergraduates, was provided to the students. Each self-selected team of students chose to read one of the following:

Arenas, A. (1995). Sum of three squares, from Gauss to modular forms. Preprint.

Davenport, H., and Watson, G. L. (1954). The minimal points of a positive definite quadratic form. *Mathematika* **1**: 14–17.

Ribenboim, P. (1989). Gauss and the class number problem. *Proc. International Symposium on Math. and Theoretical Physics, Guaruja*: 3–63.

Gerstein, L. J. (1991). Stretching and welding indecomposable quadratic forms. *J. Number Theory* **37**: 146–151.

At the end of the third and fifth weeks of the program, each student presented a progress report on her paper. My role was to direct the students to appropriate resources, and to suggest relevant questions they could answer. In the last two meetings of the seminar, we ran a mini-conference where students gave final presentations on their papers.

Assessment

I was very impressed with the students' performance. Throughout the six weeks they worked very well together. They were initially timid about presenting their solutions, but their confidence quickly grew. Their presentations were organized, clear, correct, and very nicely presented. They rarely asked for help on the exercises. I discussed their papers with them several times; their questions were always perceptive. They did a lot of independent library work to understand fully the contents of the papers they studied. Their final presentations were very professional, and, mathematically, they were well into the graduate level.

The students expressed the importance of this program to them as young mathematicians. They learned critical skills that they will need in graduate school. They found the confidence that comes from facing a strenuous, often frightening, situation and succeeding. They formed strong friendships with other young female mathematicians. They gained strength to help them continue pursuing mathematics in a climate that is often discouraging to women.

Exercise Set A

Let p be a prime. Then $\{1, 2, \ldots, p-1\}$ is a group under multiplication modulo p. Really, $\{1, 2, \ldots, p-1\}$ is a complete set of representatives for the cosets in the multiplicative group $(\mathbb{Z}/p\mathbb{Z})^\times$. That is,

$$(\mathbb{Z}/p\mathbb{Z})^\times = \{1 + p\mathbb{Z}, 2 + p\mathbb{Z}, \ldots, (p-1) + p\mathbb{Z}\},$$

which is a group under multiplication.

Lagrange's Theorem states that given a finite group G of order n, $g^n = e$ for every $g \in G$, where e denotes the identity element of G. Thus given any g in $\{1, 2, \ldots, p-1\}$, $g^{p-1} \equiv 1 \pmod{p}$. Now, for any integer a not divisible by p, $a \equiv g \pmod{p}$ for some g in $\{1, 2, \ldots, p-1\}$ and therefore $a^{p-1} \equiv 1 \pmod{p}$. Hence for any $a \in \mathbb{Z}$, $a^p \equiv a \pmod{p}$. This result is known as Fermat's Little Theorem.

Suppose a is an integer with p not dividing a. We say a is a quadratic residue modulo p if $a \equiv b^2 \pmod{p}$ for some $b \in \mathbb{Z}$. We define the Legendre symbol $\left(\frac{\cdot}{p}\right)$ by

$$\left(\frac{a}{p}\right) = \begin{array}{ll} 1 & \text{if } p \text{ \textit{does not divide} } a \text{ and } a \\ & \text{is a quadratic residue mod } p \\ -1 & \text{if } p \text{ does not divide } a \text{ and } a \\ & \text{is not a quadratic residue mod } p \\ 0 & \text{if } p \mid a. \end{array}$$

For the exercises you will need the following theorem.

1st **Isomorphism Theorem.** Suppose $\varphi : G \to G'$ is a group homomorphism. Then $G/\ker\varphi \approx \varphi(G)$. Note that $\varphi(G)$ is also denoted $\mathrm{Im}\varphi$.

Given a polynomial $f(X)$ with integer coefficients, $f(X) \equiv 0 \pmod{p}$ has at most $\deg f$ solutions X modulo p. Thus, if p is odd and $\left(\frac{a}{p}\right) = 1$ then, working modulo p, there are only two ways to write a as a square mod p. For instance, $\left(\frac{2}{7}\right) = 1$, and

$$2 \equiv 3^2 \equiv (-3)^2 \pmod{7}.$$

We also have $2 \equiv (10)^2 \pmod{7}$, but $10 \equiv 3 \pmod{7}$ so 10 and 3 are considered to be the same solution.

To solve an exercise below, you may need to use the results of previous problems.

In what follows, we identify elements of $(\mathbb{Z}/p\mathbb{Z})^\times$ with their coset representatives. So, the coset $1 + p\mathbb{Z}$ is identified with 1. Let p be an odd prime in the following exercises.

1. Define $\varphi : (\mathbb{Z}/p\mathbb{Z})^\times \to (\mathbb{Z}/p\mathbb{Z})^x$ by $\varphi(a) = a^2$. Show that φ is a group homomorphism with kernel

$\{1, p-1\}$. Conclude that exactly half the elements of $(\mathbb{Z}/p\mathbb{Z})^\times$ are squares in $(\mathbb{Z}/p\mathbb{Z})^\times$, i.e. they are quadratic residues modulo p.

2. Show that if $a, b \in \mathbb{Z}$ with $\left(\frac{a}{p}\right) = 1$ then $\left(\frac{ab}{p}\right) = \left(\frac{b}{p}\right)$. Don't forget to consider the case $p \mid b$.

3. Now suppose $c \in \mathbb{Z}$ with $\left(\frac{c}{p}\right) = -1$. Let

$$S = \{c, 2c, \ldots, (p-1)c\}.$$

Argue that S is a complete set of representatives for $(\mathbb{Z}/p\mathbb{Z})^\times$. That is, show that each of the elements in S is congruent mod p to exactly one element from $\{1, 2, \ldots, p-1\}$. Remember: $(\mathbb{Z}/p\mathbb{Z})^\times$ is a group! Deduce that for any $b \in \mathbb{Z}$,

$$\left(\frac{bc}{p}\right) = -\left(\frac{b}{p}\right).$$

4. Conclude that $\left(\frac{\cdot}{p}\right)$ is completely multiplicative; that is, for any $a, b \in \mathbb{Z}$,

$$\left(\frac{ab}{p}\right) = \left(\frac{a}{p}\right)\left(\frac{b}{p}\right).$$

So,

$$\left(\frac{\cdot}{p}\right) : (\mathbb{Z}/p\mathbb{Z})^\times \to \{\pm 1\}$$

is a group homomorphism.

5. Suppose $\left(\frac{-1}{p}\right) = 1$. Use Legrange's Theorem to show $p \equiv 1 \pmod{4}$.

6. Set $k = \frac{p-1}{2}$, and define ψ on $(\mathbb{Z}/p\mathbb{Z})^x$ by $\psi(a) = a^k$. Show that $|\mathrm{Im}\psi| \geq 2$.

7. Prove that we can choose $b \in \mathrm{Im}\psi$ such that $b \not\equiv 1 \pmod{p}$. Show that $b^2 \equiv 1 \pmod{p}$ and deduce that $b \equiv -1 \pmod{p}$.

8. Show that if $p \equiv 1 \pmod{4}$, then $\left(\frac{-1}{p}\right) = 1$.

9. Use congruence modulo 4 to show that if $p \equiv 3 \pmod{4}$, then p is not the sum of two squares (of integers, of course).

For Exercises 10 to 14, assume $p \equiv 1 \pmod{4}$.

10. Show that for some $m, z \in \mathbb{Z}_+$,

$$mp = z^2 + 1.$$

Furthermore, we can take

$$-\frac{p}{2} < z < \frac{p}{2},$$

and so $m < p$.

11. From Exercise 10, we have that for some $m, x, y \in \mathbb{Z}_+$, with $m < p$,

$$mp = x^2 + y^2.$$

This is actually weaker than the result of Exercise 10, but a useful formulation for Exercise 14. Choose $u, v \in \mathbb{Z}$ such that

$$-\frac{m}{2} \le u, v \le \frac{m}{2}, \quad u \equiv x \pmod{m},$$

and $v \equiv y \pmod{m}$. Show that

$$u^2 + v^2 \equiv 0 \pmod{m},$$

and thus

$$mr = u^2 + v^2,$$

for some $r \in \mathbb{Z}_+$ with $r < m$.

12. Use Exercise 11 to show $m^2 rp = a^2 + b^2$, where $a = xu + yv$ and $b = xv - yu$. Prove $m \mid a$ and $m \mid b$. Then use induction to show p is the sum of two squares.

For Exercises 13 and 14, suppose now that

$$p = x^2 + y^2 = X^2 + Y^2,$$

where $x, y, X, Y \in \mathbb{Z}$ with $1 \le x \le y$ and $1 \le X \le Y$.

13. First consider solutions h to the equation

$$h^2 + 1 \equiv 0 \pmod{p}.$$

Show that for some positive $h \in \mathbb{Z}$, we have

$$x \equiv hy \pmod{p}$$

and

$$X \equiv hY \pmod{p}.$$

This is the same h in the last two congruences.

14. As in Exercise 12, factor

$$p^2 = (x^2 + y^2)(X^2 + Y^2)$$

as a sum of two squares. Then show that we must have $x = X$ and $y = Y$.

For further enquiry: Several mathematicians have produced algorithms for constructing the numbers x, y such that $p = x^2 + y^2$. What are these algorithms, why do they work, and how easy are they to apply in practice?

Exercise Set B

Let p be a prime such that $p \equiv 3 \pmod 4$.

1. Consider the set $\{1, 2, \ldots, p-1\}$, and let N be the smallest element in this set that is **not** a quadratic residue modulo p. Show that

$$x^2 + 1 = -y^2 \pmod{p}$$

has a solution.

2. Show that we can choose $x, y \in \mathbb{Z}$ such that $0 < x < \frac{1}{2}p, \ 0 < y < \frac{1}{2}p$, and

$$x^2 + y^2 + 1 = mp,$$

for some $m < p$. Note that we automatically have $0 < m$.

3. Suppose

$$mp = a^2 + b^2 + c^2 + d^2,$$

for some $m, a, b, c, d \in \mathbb{Z}, 0 < m < p$. Take $A, B, C, D \in \mathbb{Z}$ such that

$$A \equiv a \pmod{m}, \ -\tfrac{1}{2}m < A \le \tfrac{1}{2}m,$$
$$B \equiv b \pmod{m}, \ -\tfrac{1}{2}m < B \le \tfrac{1}{2}m,$$

$$C \equiv c \pmod{m}, \ -\tfrac{1}{2}m < C \le \tfrac{1}{2}m,$$
$$D \equiv d \pmod{m}, \ -\tfrac{1}{2}m < D \le \tfrac{1}{2}m.$$

Show that

$$A^2 + B^2 + C^2 + D^2 = mr,$$

for some $r \in \mathbb{Z}, 0 \le r \le m$.

4. Use the notation in Exercise 3 to show that if $r = 0$ or $r = m$, then m divides p. Conclude that $0 < r < m$.

5. Show that

$$m^2 rp = \alpha^2 + \beta^2 + \gamma^2 + \delta^2,$$

where m, r, p are as above, $\alpha, \beta, \gamma, \delta \in \mathbb{Z}$, and

$$\alpha \equiv \beta \equiv \gamma \equiv \delta \equiv 0 \pmod{m}.$$

Conclude that rp is the sum of four squares.

6. Verify that for any $a, b, c, d, A, B, C, D \in \mathbb{Z}$,

$$
\begin{aligned}
(a^2 + b^2 &+ c^2 + d^2)(A^2 + B^2 + C^2 + D^2) \\
&= (aA + bB + cC + dD)^2 \\
&+ (aB - bA - cD + dC)^2 \\
&+ (aC + bD - cA - dB)^2 \\
&+ (aD - bC + cB - dA)^2.
\end{aligned}
$$

Conclude that every positive integer is the sum of four squares.

Exercise Set C

A finite continued fraction is a fraction of the form

$$
a_0 + \cfrac{1}{a_1 + \cfrac{1}{a_2 + \cdots + \cfrac{1}{a_N}}}
$$

We denote the above fraction by $[a_0, a_1, \ldots, a_N]$. The numbers $a_0, a_1, \ldots, a_N$ are called the partial quotients, or simply the quotients. This terminology comes from applying Euclid's algorithm to obtain a continued fraction for a number.

For example, to obtain a continued fraction for $\frac{67}{24}$, we observe:

$$
\begin{aligned}
67 &= 2 * 24 + 19, \\
24 &= 1 * 19 + 5, \\
19 &= 3 * 5 + 4, \\
5 &= 1 * 4 + 1.
\end{aligned}
$$

Thus $\frac{67}{24} = 2 + \frac{19}{24}$, $\frac{24}{19} = 1 + \frac{5}{19}$, $\frac{19}{5} = 3 + \frac{4}{5}$, and $\frac{5}{4} = 1 + \frac{1}{4}$.

Make substitutions to find that the so-called "complete quotient" $\frac{67}{24}$ is equal to the continued fraction

$$
2 + \cfrac{1}{1 + \cfrac{1}{3 + \cfrac{1}{1 + \cfrac{1}{4}}}} = [2, 1, 3, 1, 4]
$$

1. Find a continued fraction for $\dfrac{23}{17}$.

2. Find the complete quotient equal to $[1, 3, 5, 3, 1]$.

Observe that Euclid's algorithm allows us to find a continued fraction for each rational number larger than 1.

Given $x = [a_0, \ldots, a_N]$ and $0 \le n \le N$, we say $x_n = [a_0, \ldots, a_n]$ is the nth partial convergent to x.

From now on assume $N > 0$. Define p_n, q_n recursively, for $2 \le n \le N$:

$$
\begin{aligned}
p_0 &= a_0, \\
p_1 &= a_1 a_0 + 1, \ldots \\
p_n &= a_n p_{n-1} + p_{n-2}; \\
q_0 &= 1, \\
q_1 &= a_1, \ldots \\
q_n &= a_n q_{n-1} + q_{n-2}.
\end{aligned}
$$

Note that p_n, q_n depend only on $a_0, \ldots, a_n$, and that

$$
x_0 = \frac{p_0}{q_0} \text{ and } x_1 = \frac{p_1}{q_1}.
$$

3. Note that

$$
x_n = [a_0, \ldots, a_n] = \left[a_0, \ldots, a_{n-2}, a_{n-1} + \frac{1}{a_n} \right].
$$

Use induction to show

$$
x_n = \frac{p_n}{q_n},
$$

for $2 \le n \le N$. Then verify that

$$
x_n = \frac{a_n p_{n-1} + p_{n-2}}{a_n q_{n-1} + q_{n-2}},
$$

for $2 \le n \le N$.

4. Show that for $1 \le n \le N$,

$$
p_n q_{n-1} - p_{n-1} q_n = p_{n-2} q_{n-1} - p_{n-1} q_{n-2},
$$

and use induction to prove that

$$
p_n q_{n-1} - p_{n-1} q_n = (-1)^{n-1}.
$$

5. Show that for $2 \le n \le N$,

$$
p_n q_{n-2} - p_{n-2} q_n = (-1)^n a_n.
$$

For the rest of the exercises, suppose $a_n > 0$ for $n \ge 1$.

6. Show that for $1 < n \le N$,

$$
[a_0, \ldots, a_n] = [a_0, [a_1, \ldots, a_n]],
$$

and more generally, for $1 \leq m < n \leq N$,

$$[a_0, \ldots, a_n] = [a_0, \ldots, a_{m-1}, [a_m, \ldots, a_n]].$$

Then show that

$$x = x_N = \frac{a'_n p_{n-1} + p_{n-2}}{a'_n q_{n-1} + q_{n-2}},$$

where $a'_n = [a_n, \ldots, a_N]$.

7. Show that for $2 \leq 2n+2 \leq N$, $x_{2n+2} > x_{2n}$. Also show that for $3 \leq 2m+1 \leq N$, $x_{2m+1} < x_{2m-1}$.

8. Suppose that for some m, n with $2n$, $2m+1$ between 0 and N, we have $x_{2n} \geq x_{2m+1}$. Derive a contradiction.

9. Prove that for $0 \leq 2n < N$, $x_{2n} < x$. Also prove that for $1 \leq 2m+1 < N$, $x_{2m+1} > x$.

Now suppose $a_0, \ldots, a_N \in \mathbb{Z}$. Note that this implies $p_n, q_n \in \mathbb{Z}$ for $0 \leq n \leq N$.

10. Show that $q_0 \leq q_1$, and for $2 \leq n \leq N$, $q_{n-1} < q_n$. Also show that $q_2 \geq 2$, and for $3 \leq n \leq N$, $q_n > n$.

11. Show that $\gcd(p_n, q_n) = 1$ for $0 \leq n \leq N$.

12. Set $q'_1 = a'_1$ and for $1 < n \leq N$, set $q'_n = a'_n q_{n-1} + q_{n-2}$. For $1 \leq n \leq N$, show

$$x - \frac{p_n}{q_n} = \frac{(-1)^n}{q_n q'_{n+1}}.$$

13. Argue that for $0 \leq n < N$,

$$a_n < a'_n < a_n + 1.$$

Then show that $q_1 = a_1 < a'_1 < a_1 + 1 \leq q_2$, and for $1 \leq n \leq N-2$,

$$q_{n+1} < q'_{n+1} < q_{n+2}.$$

Hint: Rewrite q_{n+1} in terms of a_{n+1}, q_n and q_{n-1}, and rewrite q'_{n+1} in terms of a'_{n+1}, q_n and q_{n-1}. Conclude that for $1 \leq n \leq N-2$,

$$\frac{1}{q_{n+2}} < |p_n - q_n x| < \frac{1}{q_{n+1}},$$

$$|p_{N-1} - q_{N-1} x| = \frac{1}{q_N},$$

and

$$p_N - q_N x = 0.$$

14. Prove that $|p_n - q_n x|$ decreases and q_n increases, as n increases. Conclude that $|x - \frac{p_n}{q_n}|$ decreases as n increases.

15. Argue that

$$q_n x - p_n = \frac{(-1)^n \delta_n}{q_{n+1}},$$

where $0 < \delta_n < 1$ for $1 \leq n \leq N-2$, and $\delta_{N-1} = 1$. Conclude that

$$\left| x - \frac{p_n}{q_n} \right| < \frac{1}{q_n q_{n+1}} < \frac{1}{q_n^2},$$

for $n \leq N - 2$.

Exercise Set D

Suppose $a_0 \in \mathbb{Z}$, $a_1, a_2, a_3, \ldots \in \mathbb{Z}_+$. Set $x_n = [a_0, a_1, \ldots, a_n]$ for $n \geq 0$. Write $x_n = \frac{p_n}{q_n}$ with p_n, q_n as in the preceding exercises.

1. Show that the even convergents x_{2n} are bounded above; conclude the sequence (x_{2n}) converges to some $y_0 \in \mathbb{R}$. Similarly, show that the odd convergents x_{2n+1} are bounded below; conclude the sequence (x_{2n+1}) converges to some $y_1 \in \mathbb{R}$. Finally, show $y_0 \leq y_1$.

2. Show that for $n \geq 0$,

$$\left| \frac{p_{2n}}{q_{2n}} - \frac{p_{2n-1}}{q_{2n-1}} \right| \leq \frac{1}{2n(2n-1)},$$

and conclude $y_0 = y_1$.

Thus we set $x = \lim_{n \to \infty} x_n$, and we write $x = [a_0, a_1, a_2, \ldots]$.

Now fix $\alpha \in \mathbb{R}, \alpha \notin \mathbb{Q}$. Take $a_0 \in \mathbb{Z}, \alpha_1 \in \mathbb{R}$ such that

$$\alpha = a_0 + \frac{1}{\alpha_1}, \qquad 0 < \frac{1}{\alpha_1} < 1.$$

For $n \geq 1$, take $a_n \in \mathbb{Z}_+$, $\alpha_{n+1} \in \mathbb{R}$ such that

$$\alpha_n = a_n + \frac{1}{\alpha_{n+1}}, \qquad 0 < \frac{1}{\alpha_{n+1}} < 1.$$

3. Show that it is possible to choose a_n, α_n as described, yielding an infinite sequence $a_0, a_1, a_2, \ldots$ ($a_n > 0$ for $n \geq 1$). Also show that $\alpha_n \notin \mathbb{Q}$ for $n \geq 1$.

For $n \geq 0$, set $x_n = [a_0, a_1, \ldots, a_n]$ and define p_n, q_n as in Exercise Set C.

4. Prove that

$$\alpha = \frac{\alpha_{n+1} p_n + p_{n-1}}{\alpha_{n+1} q_n + q_{n-1}},$$

for $n \geq 1$.

5. Show that

$$\alpha - \frac{p_n}{q_n} = \frac{\pm 1}{q_n(\alpha_{n+1} q_n + q_{n-1})}.$$

6. Show that

$$\left| \alpha - \frac{p_n}{q_n} \right| < \frac{1}{q_n q_{n+1}},$$

and conclude that $\frac{p_n}{q_n} \to \alpha$ as $n \to \infty$.

Here is an example of how to construct an infinite continued fraction. Take $\alpha = \sqrt{2}$. Since $1 < \alpha < 2$, $a_0 = 1$ and $\frac{1}{\alpha_1} = \sqrt{2} - 1$. Then

$$\alpha_1 = \frac{1}{\sqrt{2} - 1} = \sqrt{2} + 1 = 2 + \frac{1}{\alpha_1}.$$

So, $a_1 = 2$ and $\alpha_2 = \alpha_1$. Thus for $n \geq 2$, $a_n = 2$ and $\alpha_{n+1} = \alpha_1$. To prove this formally, we would use induction. Consequently, $\alpha = \sqrt{2} = [1, 2, 2, 2, \ldots]$, which is abbreviated $[1, \dot{2}]$.

7. Find a continued fraction for $\sqrt{3}$.

8. Find a continued fraction for $\sqrt{5}$.

9. Find a continued fraction for $\sqrt{6}$.

Exercise Set E

We write

$$[a_0, \ldots, \dot{a}_m, \ldots, \dot{a}_{m+k}]$$

to denote the "periodic" continued fraction

$$[a_0, a_1, a_2, \ldots],$$

where $a_n = a_{n-k-1}$ for $n > m + k$.

It is a fact that for $N \in \mathbb{Z}_+$ and N not a square, we have $\sqrt{N} = [a_0, a_1, a_2, \ldots]$. Then

$$\sqrt{N} + a_0 = [\dot{2}a_0, a_1, \ldots, \dot{a}_k],$$

for some $k \geq 0$, i.e., the continued fraction for $\sqrt{N} + a_0$ is "purely periodic".

1. With notation as above, show that

$$\sqrt{N} = \frac{\alpha p_k + p_{k-1}}{\alpha q_k + q_{k-1}},$$

where $\frac{p_k}{q_k} = [a_0, \ldots, a_k]$, $\frac{p_{k-1}}{q_{k-1}} = [a_0, \ldots, a_{k-1}]$, and $\alpha = [\dot{2}a_0, a_1, \ldots, \dot{a}_k]$. Deduce that

$$\sqrt{N}(\sqrt{N} + a_0)q_k + \sqrt{N}q_k = (\sqrt{N} + a_0)p_k + p_{k-1}.$$

2. Show that $Nq_k = a_0 p_k + p_{k-1}$ and that $a_0 q_k + q_{k-1} = p_k$. Deduce that

$$p_k(p_k - a_0 q_k) - q_k(Nq_k - a_0 p_k) = (-1)^{k-1}.$$

3. First suppose k is odd. Argue that $x = p_k, y = q_k$ yields a solution to Pell's equation $x^2 - Ny^2 = 1$. Describe how to obtain a solution to Pell's equation when k is even.

Model
Programs

Encouraging Women in Mathematics:
The Spelman-Bryn Mawr Mathematics Programs

SYLVIA T. BOZEMAN AND RHONDA J. HUGHES

Spelman College and Bryn Mawr College

The Mathematics Departments of Spelman College, Atlanta, Georgia, and Bryn Mawr College, Bryn Mawr, Pennsylvania, joined their efforts to operate mathematics enrichment programs designed to identify and encourage talented college women to pursue careers in the mathematical sciences. The primary activities of the program were research-based Summer Mathematics Programs conducted from 1992 to 1995. Aimed at women who had just completed their freshman or sophomore year, the programs enrolled, in the four-year period, a total of 33 undergraduate women from colleges in the areas nearby the host institutions. These programs both extended and complemented mentoring activities at the two colleges by offering an introduction to mathematical research with emphasis on developing skills in the use of computers as research tools.

Each year, the program was planned and coordinated by Rhonda Hughes and Sylvia Bozeman, mathematics professors at these two women's colleges with reputations for having large numbers of mathematics majors. While they were each serving as chairperson of their respective department, the two discovered their mutual concern over the progress of women and minority students at advanced levels in mathematics. During the first two years the Summer Program was conducted at Bryn Mawr College, and named the Bryn Mawr–Spelman Summer Mathematics Program; in the final two years it was hosted at Spelman College, and named the Spelman–Bryn Mawr Summer Mathematics Program. Thus, public information on the programs, even within the National Science Foundation, the funding agency, might exist under either or both captions.

Goals

The eight-week program was based on the belief that specific steps can be taken to produce students who are more self-confident, more aware of mathematics as a field to which they might enjoy contributing, better able to see themselves as future mathematicians, and more committed to a career in teaching or research, thereby increasing the number of women who choose mathematical careers. The steps to be taken include:

- exposing students to mathematical research;

- encouraging students to work in teams and share ideas;

- providing students with opportunities to present their work; and

- having young women meet successful female and minority mathematicians.

All of the steps were incorporated into the components and activities of the programs.

The Research Projects

Each summer several faculty members and a few graduate students were recruited to work with students in the program. Several topics were presented during the first week, and the students were given an opportunity to choose the project on which they wished to work for the remainder of the program. Research was conducted by teams of students supervised by local college and university faculty or by women graduate students who were recruited to serve as graduate assistants and mentors. In addition, students received instruction in the use of Mathematica or Maple, both computer algebra systems. Dr. Anthony Hughes at Bryn Mawr College, who provided computer instruction, and Dr. Fred Bowers at Spelman College, who provided computer instruction and coordinated the academic activities, made major contributions to the programs. Much of the work was computational in nature, although theoretical aspects of the topics were studied as well. All work at Bryn Mawr was done in the Julia Martin Cheever Computer Classroom, and all work at Spelman was done in the Electronic Classroom; both classrooms were established with earlier partial support from National Science Foundation grants. Particularly successful was the move to offer the graduate student mentors, who were sufficiently advanced, the opportunity to direct projects under additional supervision by faculty. The graduate students also offered mini-courses or tutorials-on-demand in topics such as linear algebra, proof techniques, abstract algebra, and topology.

Student research projects over the four summers included wavelets and Mallat's algorithm, graph theory, minimal surfaces, in particular the Costa surface, number theory, error-correcting codes, random number generators, and elliptic curves. The minimal surface project, done in the second year under the direction of Professor Victor Donnay of Bryn Mawr, resulted in a video of a Mathematica animation of the construction of the Costa minimal surface from a torus. A later version of this video, completed by Donnay and his students in the Geometry Center at the University of Minnesota, became part of an exhibit at the Maryland Science Museum.

Mentoring Activities

Considerable emphasis was placed on promoting interactions between the undergraduate and graduate women, as the graduate mentors served as important role models for the undergraduates. Housed near the participants, the graduate students provided assistance during the daytime and the evenings, and joined in the cookouts, trips, and other social outings. Their presence added significantly to the value of the experience and, in the opinion of many participants, was one of its finest components.

In order to provide additional mathematical exposure, weekly guest lecturers were invited to present work on a mathematical topic, to discuss career options, and to interact with students informally. Several of these were female or minority mathematicians. Field trips to research facilities at telecommunications and other types of industrial sites included conversations with scientists there. Among the longer-term visitors was Dr. Danielle Carr, then a postdoctoral mathematical biologist at New York University, who spent two weeks with the participants. Dr. Carr gave several talks on her work and her experiences as a minority female mathematician. The students seemed to appreciate these glimpses into the lives of mathematicians and other scientists.

A special effort was made to help students develop their self-confidence and skill in giving oral presentations. Each week the students gave updates on their work, supporting one another, applauding each other's successes, and offering encouragement when needed. Through supplemental funds provided by Spelman and Bryn Mawr Colleges from outside sources, all participants attended the national Joint Mathematics Meetings in the January immediately following the Summer Program, and most of them gave presentations of their work there.

Program Impact

Requirements which resulted in the selection of a diverse group of participants were built into the program. In the last two years, 59% of the women participants were from groups underrepresented in the sciences and mathematics. In addition to students from Spelman and Bryn Mawr Colleges, program participants included students from Clark Atlanta University, Morris Brown College, Haverford College, Swarthmore College, Community College of Philadelphia, and Kennesaw College. With this diversity, participants learned to appreciate other backgrounds and cultures, to work on diverse teams, and to support each other in their investigations. Several criteria were used in measuring the impact of the program. Immediate impact was assessed through an essay, which each student wrote at the end of the summer reflecting on the summer's experiences, and through the activities of the student during the subsequent summers. Longer range impact was determined by the student's persistence in mathematics or in a scientific area, and by the student's decision regarding graduate school.

The students from these programs are targeted to remain in the mathematical sciences and earn graduate degrees in these areas. To date, three of the program's participants have been awarded Goldwater Fellowships, and one received a National Science Foundation Graduate Fellowship. Of the 16 participants from the 1992 and 1993 programs, all subsequently attended Research Experiences for Undergraduates (REU) programs or held summer internships. Fourteen have earned bachelors degrees in mathematics, and nine of those have entered graduate programs in the mathematical or computer sciences. Of the 17 student participants for 1994 and 1995, 10 entered programs that provided research experiences in the following summer. Seven have now earned a bachelors degree in mathematics. Of those, three entered graduate programs in mathematical areas in 1996, and another expects to enter in June 1997. Ten of the students are still at the undergraduate level. Those who have entered graduate programs in the mathematical sciences, including computer science, are at Clark Atlanta University, Columbia University, Johns Hopkins University, Massachusetts Institute of Technology, North Carolina State University, Rice University, Syracuse University, and University of Oregon; one is in physics at the California Institute of Technology. One recently completed her PhD, and obtained an academic position.

There are two areas where we are not able to assess the impact—the impact on the graduate assistants and the impact of the dissemination effort. Although it was clear that they enjoyed the experience of mentoring younger students, one goal was to provide inspiration and support to the graduate assistants in their own graduate studies. Six of the seven graduate students have continued to pursue their studies at Columbia University, Rutgers University, State University of New York at Stony Brook, University of Maryland, Bryn Mawr College, and Georgia Institute of Technology.

The programs have been discussed widely by the two Principal Investigators, including presentations at the annual programs of some scientific societies. Articles devoted in whole or in part to a discussion of the program have appeared in several newsletters of professional mathematics societies, such as the Association for Women in Mathematics, and the Society for Industrial and Applied Mathematics. Tara Brendle, both a second-year participant and a fourth-year graduate assistant, gave a poster session on the program at the International Women's Conference in Beijing, China, in September 1995. The student participants advertised the program when they gave poster presentations on their research at the four Joint Mathematics meetings which they attended. In addition, several also have given presentations in their local areas. We hope that the dissemination of information about these programs has also had some impact on the actions of others in the mathematics community by increasing support for the development of women and minority students.

The University of Michigan REU[1] Program in Mathematics

DANIEL M. BURNS, JR. AND DONALD J. LEWIS

University of Michigan

The Mathematics Department at the University of Michigan has for nine years conducted a research program for its majors and other students, funded mainly by supplements to National Science Foundation (NSF) research grants, but also by department endowment funds. It is further supported by the University, which forgoes overhead recovery on any REU funds coming into the department because the basic principles of the NSF's REU program match so closely the University's own attitudes towards undergraduate research experiences. The number of undergraduate participants in our program has varied widely over the years, from four or five per year in the beginning to a high point of 26 per year in 1994. In 1996-1997 we had 11 completed projects. On average about 25-30% of the participants have been women.

Students make application to the program and ask faculty familiar with their work to recommend them. The students typically have completed, at the very minimum, four courses beyond sophomore calculus, although other useful elements in their background may compensate for less coursework preparation, depending on the specific faculty projects available in a given year. They also should have shown initiative and be reasonably self-directed. In pairing students and faculty, the ideal is always that the student should already have expressed an interest in such work to a faculty member with whom the student has had a course. Faculty announce the availability of the program to their classes during the previous academic year. Most students still do not present themselves to a potential faculty mentor in such a direct way, however, and we invite students to apply to the program in general and we try to place students who seem qualified with suitable mentors. We have the student write of his or her mathematical interests in the most extensive terms possible so that the program director has ample possibilities for finding connections between the student's interests and the interests expressed by the faculty that year,

who have been canvassed independently by the program director. The program director then consults with faculty members on their interest in a number of seemingly relevant applicants. If a match seems possible, the program director passes the information back to the student applicant, who is expected to set up an appointment with the prospective mentor to see whether things can indeed be worked out to mutual advantage. Sometimes students have several mentors to choose from, and sometimes faculty may have an overabundance of candidates. The faculty sponsor and the student meet several times to decide whether the professor's research problems interest the student, whether the problem is at a level such that the student is likely to be able to make a contribution, and whether the student and professor are compatible.

The research advisors come primarily from the Mathematics Department faculty, but we also involve faculty with strong mathematical interests from other units on campus such as statistics, biology, medicine, chemistry, economics, computer science, physics, and operations research. The projects are all mathematical in nature. The choice of the faculty participants varies from year to year, depending on who has a suitable problem and the time to work with a student. Most of the faculty participants have NSF or Department of Defense grants and are well-established researchers, and a few may be postdoctoral faculty who may have NSF or Michigan funding. The faculty are not explicitly compensated for this contribution. Many are supported during this period from their research grants, the output of which is enhanced by the students' work. Some of the faculty mentors are not compensated at all for this, but simply contribute their time because they find it satisfying.

Indeed, this "in and out" participation of our faculty is important for the long-term sustainability of a large-scale effort like this. No one faculty person has a burdensome responsibility for the continuation of the program, and faculty participate when they have fresh questions to propose. It should be mentioned that this is one aspect of the program where the biggest change has occurred: initially it was very difficult to find enthusiastic faculty to be mentors. With time, however, the success of the first students changed the minds of many of the skeptics, and gradually the culture of the department has changed. Now faculty mentoring of undergraduate research is considered a natural thing to do, and faculty are much more comfortable doing this. They realize more concretely that there are, indeed, things which

---

[1] Research Experiences for Undergraduates

many of these students can do which are of real scientific interest to faculty. In addition, they now think ahead with this in mind, and keep their eyes open for possible problems or research areas which a good undergraduate could use. It is now easy to find many more enthusiastic faculty mentors than are needed in any given summer.

The problems the students investigate are mainstream and of interest to the research community. The problems are usually from some portion of the professor's ongoing research program—a portion that is at the level where a student can make a genuine contribution, which is reasonably self-contained, and which can lead to some conclusions with eight weeks' effort, the length of time the students will be supported on this program. Sometimes the project is from a new area a faculty member is trying to explore, and in such circumstances the faculty "learner" is even closer to the level of the student participant. While all the projects are mathematical in nature, many, but certainly not all, involve computation and experimentation as a means to test or formulate conjectures. The students sometimes spot patterns a sophisticated eye overlooks!

The students work one-on-one with a faculty advisor or in small teams of two or three. Initially each professor meets with that professor's students daily. As the students begin to comprehend the problem and gain confidence, the meetings with the professor may well diminish to twice a week. Some years the students have shared their experiences, frustrations, and joys at a weekly student-run seminar attended by all participants at which the students report on their projects and progress. At the seminar, students give each other encouragement and suggestions as to possible next steps. As a concluding element of the project, the students write a report on what they have accomplished. The writing experience is considered an essential element of the research experience.

There are occasional social functions for the participants, but these are fewer than those in REU site programs, since the student participants all live in their separate apartments off campus which they have had for the school year.

Although the students are funded for an eight-week period, frequently they work longer. Some continue to work on the project during the ensuing academic year. A few of our students, normally two or three, are sufficiently successful that their work leads to a publication with them as sole author or possibly joint with the professor. Others have their work noted by a footnote in a subsequent publication by the professor. On the whole, faculty have been quite generous in acknowledging the students' contribution. More common is for a student participant to report on their work and experiences in the Undergrad Math Society's seminar the following academic year. In general, we keep a (now large) book of student reports on their findings or attempts in the various projects, and pass this book along to succeeding years' students to show them what they could hope to accomplish.

To compare this program to the REU site programs offered around the country, it would seem that ours is better suited to a sustainable effort at a large public research university. We see our advantages being that this program is closer to a model of graduate student work and life, or real-time research in the outside world. Students work less in focused groups on one or two common problems. There is less emphasis on preparatory mini-courses. Since the students are already here in the spring, the faculty mentor usually has the student get a head start on things by reading well before the project actually begins. We also find that we have a good opportunity to follow-up on open ends of the project during the succeeding term, although competing commitments limit this practically. Finally, our "matching service" approach can open the broadest possible set of research area choices to the student. In this way we can use the size and research depth of such an institution maximally to the students' advantage.

We have observed that participating in the research program has generated in the women-participants considerable self-confidence in their abilities. Indeed, if the student, professor, and problem all "click", there is enormous growth over the summer. Part of this self-confidence obviously comes from experiencing success and some comes from seeing that they can do as well as others regardless of gender. Part of it also comes from the fact that the students see a large number of women actually engaged in high-level research. In any given year, there are 10 to 12 women amongst the Department's tenured, tenure-track, and three-year assistant professors. All are actively engaged in research, and many are involved in the summer research program. Also, there are approximately 20 women graduate students who are working on their theses during the program session. These women faculty and graduate students typically form a support group and information source for the women participants. This is facilitated by a mathematics department staff person who takes a personal interest in encouraging these students.

From our studies of women graduate students in mathematics and physics, we have found that nearly all had a mentor who played a significant role in the development of these students. The mentor, who can be of either sex, encourages, maybe pushes at times, is concerned for their welfare, and remains constant in support as they progress to the doctorate and beyond. Our experience suggests that the mutual respect and support that develop between the junior and senior researcher during these summer experiences do lead to this type of mentoring. A large percentage of the participants, both men and women, go on to graduate studies, though this may be in a variety of (usually) mathematical areas. We encourage students to consider possible application areas when setting up their summer projects, and several have followed up on this in their subsequent career plans. It seems, though this is just an impression, that many of the young women seem to discover the applied side of mathematics in these projects in a way that seems to be a bit of a revelation. It is unclear whether there is a particular impedance against this sort of interest or career path beforehand, but this *seems* at times to be an especially invigorating form of the REU experience for young women.

Up to now, all of this evidence is, unfortunately, anecdotal in nature, at least as far as what we have gathered ourselves. The last time the department was examined by an external review panel, however, they interviewed undergraduates extensively, and according to their interviews, the students seem to view the REU program at least as favorably as the faculty does, and for much the same reasons. We find this very encouraging and another good reason to continue to offer the program.

The Mills College Summer Mathematics Institute

Ani Adhikari, Steven Givant, and Deborah Nolan

University of California, Berkeley; Mills College; and University of California, Berkeley

What is the Mills Summer Mathematics Institute?

The Mills College Summer Mathematics Institute (SMI) is a six-week intensive summer mathematics program for 24 talented undergraduate women, that is funded by the National Science Foundation and the National Security Agency. The aim of the SMI is to increase the number of bright undergraduate female mathematics majors who continue on into graduate programs in the mathematical sciences and obtain advanced degrees. The goals of the program are: (1) to provide the participating students with an intensive immersion in mathematics, designed to motivate them, and to strengthen their preparation, for entrance into a PhD program in the mathematical sciences; (2) to broaden and strengthen the efforts of the nation's mathematics faculties to encourage more women to major in mathematics and to direct the most promising ones toward graduate school; (3) to encourage participating students to serve as leaders in mathematical activities among their peers at their own institution; (4) to gather and analyze data, with a view to clarifying the reasons why a much smaller fraction of women mathematics majors than of men go on to PhD programs in mathematics.

How did the Mills Summer Mathematics Institute start?

It was conceived in 1990 during a student strike. Mills had been a women's college for 138 years when the Trustees voted, in 1990, to make the College coeducational. There were strong objections by students, alumnae, and many faculty, and immediately after the announcement, the student strike began. During this period there were many discussions about the benefits of being educated at a women's college, and the difficulties that women often face, especially in science classes, at coeducational institutions.

The well-known logician and mathematics educator, Leon Henkin, was teaching at Mills College as a visiting professor that semester, and he and Steven Givant often talked about the strike and about the pros and cons of women's colleges. Towards the end of the strike, Leon mentioned a new program that he and Uri Treisman had started at Berkeley the preceding year, a Summer Mathematics Institute for Minorities. Steven's reaction was that it would be great to have a summer mathematics institute like that for women. He had learned many things about teaching mathematics to women, while working with a colleague at Mills, Lenore Blum, one of the original and most eloquent voices in the struggle to make people aware of the problems women faced at all levels of their mathematical education and their mathematical careers. Leon, Steven, Lenore, and Diane McEntyre, a professor of computer science at Mills, together designed the program, and submitted a one-year proposal to the Division of Mathematical Sciences at the NSF. The program was on its way.

Why do so few women go on in math?

According to 1992 statistics, women make up about 45% of the mathematics majors nationally, but only about 25% of the graduate students. At institutions with very strong mathematics programs, the picture is bleaker. For example, at the University of California at Berkeley about 30% of the undergraduate mathematics majors are women, and only 14% of the MAs and 9% of the PhDs in math are awarded to women.

Some of the reasons why women don't continue their studies may include the following: few math professors are women; male students tend to dominate math classes and get more attention from the instructor; there is a common view that mathematics is a male subject; and there is a lack of awareness about women mathematicians. Women studying mathematics often have to deal not only with the difficulties inherent in the subject, but with the problems caused by studying in such an environment.

While not all women students may be affected, many are. For example, the women who have participated in the Mills SMI are among the strongest undergraduate mathematics majors in the country; yet many of them report that, before entering the summer program, they had doubts about their abilities to succeed in a graduate program. It is a tragedy that some of our brightest young women are being lost to mathematics because of such factors.

Why a women's program?

It was our hope that, in a program aimed exclusively at women, it would be possible to break some of the stereotypes. We thought that the students would be excited by the idea of doing mathematics with other women, and that they would encourage each other. Thus, one goal was to bring together a critical mass of talented women math majors. We wanted more than 10, and we felt that more than 30 would be too big for the students to form close bonds in just six weeks. Thus, we eventually settled on the number 24.

We also decided right from the start that, if possible, we wanted to hire only women instructors and TAs. It wasn't that we felt men could not be effective teachers of women, but we wanted instructors and TAs who could also serve as role models for the students. We thought that role models would add an important dimension to the whole experience for the students and perhaps even for the teachers. In particular, the women instructors and teaching assistants might themselves be very excited by the idea of teaching a class of gifted women; they might feel almost a missionary sense of enthusiasm.

We hoped that a spirit of camaraderie would spring up among the students, and between the students and instructors. We wanted to provide them with an experience that they could carry back to their home institutions, an experience that would prepare them mathematically and emotionally for graduate school. We wanted to communicate to them that women can and should be doing mathematics, that they are part of the nucleus of a growing network of professional women mathematicians.

We weren't disappointed in these dreams. Virtually every student has said that she benefited from the program in some very essential way. Here are two quotes that are typical of the student responses in anonymous end-of-program evaluations.

> It was awesome! I only know of three or four women math majors at my university, so this was a real experience for

me. It made me realize that there are other talented women in math, and although at first I felt really discouraged (since I wasn't the "best"), I came to really respect these other girls. Also, I believe this is my first step in "contact building." I will definitely hold on to the list of participants in this program and will try to keep in touch with them.

> I think that the most valuable aspect of the program for me was meeting and getting to know such remarkable women math majors. It is nice to know that I'm not alone as a woman who likes math. My experience with the Mills program was perhaps the best mathematical experience of my life. Beforehand, I viewed math as one of several subjects I might pursue. Now I can hardly imagine NOT being a math major.

How is the SMI organized?

It runs for six weeks each summer. The heart of the program consists of four seminars, two in classical areas of mathematics and two in more specialized areas. Each student elects two seminars, one from each pair. To give some examples of seminar topics, in 1994 the seminar topics were: Algebraic Number Theory, Topology, Dynamical Systems, and Graph Algorithms.

The character of the seminar work is very different from that encountered in typical undergraduate mathematics courses. Students establish by themselves many of the results. They are given challenging problems to solve, and they are assigned individual and group projects for which they read journal articles, do independent research, and report on their findings to the class. Several examples of these seminars appear in the articles within Course Designs herein.

In addition to the seminars, there are mathematics colloquia that are held twice weekly. In these talks, well-known mathematicians introduce the students to a panorama of topics. For example, among the 1993 speakers, we had Richard Karp giving a talk on computers as puzzle solvers, Wu-Yi Hsiang on sphere packing and spherical geometry, David Blackwell on game theory, Florence Lin on molecular dynamics, William Thurston on self-similar tilings of the plane, and Lenore Blum on the contemporary history of women in mathematics.

There are also four panel discussions on the following topics: applying to graduate school and getting fellowships; the variety of graduate programs available in the mathematical sciences; the experiences of women in mathematics; and non-academic careers in mathematics.

What problems have we encountered?

As such problems may be encountered by other programs, we would like to report on them.

The mix of students.

We wanted to accept a broad mix of students, not just students who were headed for graduate school on their own. The students we admitted fell into three groups: 1) students with high potential who were studying at small, sometimes isolated institutions; 2) students at large regional state universities; 3) students from well-known institutions with strong mathematics programs, some of whom were already planning to enter graduate school. We also admitted students who were at widely different stages of pre-graduate studies, from a gifted high school senior to a graduating college senior.

We included students from mathematically strong, well-known institutions for several reasons. First of all, even very bright and well-prepared women often decide not to attend graduate school, or else drop out of graduate school after a year or two. Secondly, we felt that such strong students would make an important contribution to the success of the other participants by acting as role models for the less motivated or less well prepared.

Initially, the variation in backgrounds, motivation, and preparation created difficulties. Some students felt intimidated: they had always been the best among the math majors at their home institutions, and now they were working with students who seemed to know more than they did. The instructors, too, had to work very hard to handle this diversity. By the end of the program, most of the students felt that this variety of backgrounds was a strength of the program. Younger or less experienced students definitely began to see some of the older or stronger students as role models, and the stronger students viewed others as sisters whom they wanted to help along.

Still, diversity of background and level of preparation remains a difficult issue to handle, even today. In our admission process, we now ask students to submit two faculty recommendations instead of one, and to give us a list of all collegiate math courses they have taken, including the name of the text they used. But we simply can't get enough information to be able to select a uniform group of students. Maybe such a grouping is really impossible to achieve, and perhaps it is not even desirable.

The seminar atmosphere and the intensity of the program.

When the students come to a program like the Summer Mathematics Institute, they expect something different from what they are used to back home. In the very first year, during the initial weeks of the program, students complained that attending seminars was too much like going to class back home. They complained that the program was not intense enough. But when instructors began to give them more work and more responsibility, they complained that there was too much work and not enough time to do it. The whole problem of creating an intense, exciting atmosphere is a challenging one and crucial to making the program a success.

The presence of men.

In a program aimed exclusively at women, the presence of men can sometimes be a problem. For example, for two years Steven Givant directed the Mills SMI. Even though the program was going quite well, the fact that a man was the director bothered some of the students. We think that the problem of having men in positions of authority in a program for women, and the students' perception of that, is a sticky issue.

The problem of funding.

Not surprisingly, the biggest ongoing problem is funding. In 1993 and 1994, our funding was cut successively by about 20%. At that time, the point of view of the NSF was that they wanted to provide seed money for programs with new ideas. Established programs should seek funding from other sources. We did receive two generous donations from Genentech, a biotechnology company in the Bay Area, and we were able to secure supplementary grants from the National Security Agency. Since 1995, the program has been supported by yearly grants from the NSF and NSA. Also, after the program came under the aegis of the University of California at Berkeley in 1996, we received funds from the University to help cover administrative costs.

Perhaps there are untapped sources of funding out there for worthwhile programs such as our own. But it takes time, skill, and expertise to find such sources.

How does the SMI compare with REUs?

It is informative to compare the experiences of students in the SMI and in other summer math programs, in particular the REUs sponsored by the NSF. All of these programs attempt to encourage students to become research mathematicians, but they differ in some important respects.

To succeed at an REU, a student must have the confidence and mathematical maturity to tackle an unsolved problem. Experience in the SMI admissions process shows that students who are admitted to the program but opt to go to REUs are often exceptionally strong and from rigorous and selective institutions—for example, in 1996 these students came from Bryn Mawr, Harvey Mudd, MIT, Oberlin, and St. Olaf. But the nature of the SMI allows it to reach a large number of gifted students whose backgrounds are not so strong, especially those from schools which do not have high-ranking math programs.

To gather more information for a comparison, we asked past SMI participants whether they had also attended REUs or other summer math programs. Those who had were from Berkeley, Bowdoin, Brown, Chicago, Harvey Mudd, Mt. Holyoke, and North Carolina State. Some of them attended their REU before coming to the SMI, some after, and some both before and after.

These students have made candid comparisons of their experiences in the two types of programs. From their point of view, the programs were financially quite comparable.

Regarding the intellectual aspects of the REUs, some students wrote about the thrill of working on open problems, and were grateful for the early experience in research:

> ...It's an unsolved problem ... by definition, it's never been done before, so you can't just go look up a solution. Personally I love this feeling ...

Of course, research has its attendant frustrations:

> I felt like a failure because of lack of results ... I think that REUs are riskier than the SMI—since there is a definite possibility that progress will be minimal, which is disheartening. It is also unfair, since this kind of frustration is something everyone feels, but not usually with a ten-week deadline.

As has been seen repeatedly in feedback from students and faculty, a crucial feature of the SMI is that it involves only women, and provides a long-lasting support group for its participants. The following quotes capture the students' opinions on this subject:

> [The research program] was a good experience cause it helped me see if I would really enjoy doing research, but it did not give me the contacts and positive role models that the SMI did.

> I didn't like [the REU] ... Other students ... were competitive but not supportive (e.g., doing math in the evenings wouldn't make one popular) ... It was also in that program that I first realized it can be a really big disadvantage to be a woman in mathematics.

> [The research program] was not as positive because my interaction was limited to one researcher who was not cognizant of my background ... SMI was very good because of the group interaction ... it tended to give me confidence rather than destroy it.

It seems clear that REUs and the SMI both have similar overall goals, but are different in content and aimed at somewhat different groups of students. Thus they should be thought of not as competitors but as companions. In the words of a student:

Both programs have had a good effect on my intentions to go to grad school—the SMI gave me a basis for math research in general, and the REU gave me a specific field ...

Where are the students now?

Of the 24 participants in the 1991 program, one has received her PhD and is now with the NSA, and 14 are currently in graduate school: 12 in mathematics and two in statistics. Three have received Master's: one in operations research and two in statistics. (One of these students plans to enter a PhD program in Economics in Fall 1997.) Of the remaining six students, one is a high school teacher in math, one is a high school teacher in physics, two are consultants, one is developing software, and one is an accountant.

Of the 25 participants in the 1992 program, 13 are currently in graduate school: seven in mathematics, three in applied math, two in statistics, and one in operations research. Two have received Master's: one in statistics and one in math. One is in medical school, and another is in law school. Of the remaining eight students, six are working in areas of mathematics: four are teaching mathematics (two in high school, one in community college, and one in the Peace Corps) and two are accountants.

Of the 18 participants in the 1993 program, 12 are currently in graduate school. Of these, 11 are in PhD programs in math or applied math, and the 12th is in a Master's program in education. Of the remaining six students, one received her Master's degree in math and is teaching math in high school, two are teaching math in junior high schools, one is working on software development, one is working in finance, and one returns this year from three years with the Peace Corps.

Of the 25 participants in 1994, 19 are currently in graduate school: ten in PhD programs in math or applied math, three in PhD programs in statistics or biostatistics, three in Master's programs in math, and one each in chemistry, medicine, and education. Of the remaining six students, one has her Master's degree in statistics and is working for a market research firm, one obtained the Watson fellowship to study in Australia and New Zealand, and plans to be a biostatistician, two are working in nonmathematical areas, and two are still undergraduates.

Of the 20 participants in 1995, 12 entered graduate school in the fall of 1996: nine in PhD programs in math or applied math, one in a PhD program in biostatistics, one in a Master's programs in education, and one in a teaching credential program. Of the remaining eight students, five have graduated, and three of these five have definite plans to attend graduate school in 1997. The remaining three are still undergraduates, and they all plan to go on to graduate school.

What do students and faculty think about the program?

A comprehensive evaluation of the SMI was undertaken by Ani Adhikari and Deborah Nolan in the summer of 1996. The evaluation had three components:

- Surveys of the 1993 and 1994 participants, for information on the influence the program had on them.

- Surveys of the faculty who wrote letters of recommendation for students admitted to the 1994 and 1995 programs, for their perceptions of the effect the program had on the student.

- A brief survey of the graduate advisors of the 1991 and 1992 SMI participants who are now in graduate school, to see whether the student is making satisfactory progress towards her degree.

We include a summary of some of the main responses to these surveys here.

Results from the student survey.

Altogether, 80% (34 out of 43) of the students responded to the survey. The questions in the surveys for the 1993 students and the 1994 students were similar; however, the 1994 students were asked about the effect of the program on the following year of undergraduate school. Over 60% of these students said that the program had a great deal of effect on their choice of advanced undergraduate courses. (Others pointed out that their schools did not offer many advanced courses.) The students also engaged in mathematical activities outside the classroom: over half gave talks at their home institutions, and over half participated in math clubs or conferences. Students appreciated the wealth of information provided by the program on the process of applying to graduate programs and applying for fellowships.

Questions about the effect of the program on their graduate experience were asked of both 1993 and 1994 students. Students are admitted to the SMI only if they have strong math records. Thus, one might expect them to be strongly predisposed towards graduate school before coming to the program. Nevertheless, the responses to the survey make clear that the program plays an important role in shaping their decisions about graduate school.

About half of the students said the program had a great impact on their motivation to do graduate work, and about half said that it provided them with a sense of what grad school would be like.

> I feel that I was better prepared to handle the demands of graduate school ... The atmosphere of the program opened my eyes as to what would be expected of me.

The strongest influence of the program is on the students' estimation of their own capabilities. Over two-thirds reported that their self-confidence was greatly enhanced by the SMI, and these results were confirmed by the undergraduate and graduate advisors. According to one student,

> Perhaps the program's main advantage for me is that I feel very comfortable being in grad school. That is, I feel that I belong here, as opposed to some of my female peers who have many doubts about their ability and place in this environment.

Over 60% of the students strongly agreed with the statement, "My work in the program showed me I enjoyed doing challenging math." It is worth noting that two of the students who strongly disagreed also said that the program convinced them that graduate school was not their goal; this is a valuable lesson, even though in a sense it is negative. In addition, over half the students strongly agreed with the statement that the program showed them "how to learn advanced math." This percentage is surprisingly high, given that students in the program are selected for their ability to do mathematics, and it underscores the difference between work in the SMI and in standard undergraduate classes.

In end-of-program evaluations, the students have been consistently and overwhelmingly positive about the "all women" nature of the program. It is now clear from the survey that this effect is long-lasting. Over 80% of the students have stayed in touch with fellow students from the program, and over 70% with their professors in the program. Over half the students asked for letters of recommendation from their SMI professors. A student sums up the opinion of the vast majority as follows:

> Until attending the SMI, I had only had one female math professor. Ever. I think I now have a great advantage in having discovered some positive female role models in mathematics. ... I found the program ... to be extremely helpful to seeing myself as a mathematician. I don't recall ever seriously being told that girls don't do math. But on the other hand, until attending SMI, I rarely actually saw them doing it.

Results from the undergraduate faculty survey.

The response from the undergraduate faculty who wrote letters of recommendation for the SMI participants was very positive. Three quarters (33) of them responded. Six were unable to judge the effect of the program, because they had no contact with the student after her return.

For the faculty who did have contact with the student after her return, more than 80% said the program was very beneficial. Also, about half said their student's participation in the program had an effect on the whole department. Two themes recurred throughout their comments: they noticed a tremendous increase in the self-confidence and they noticed a tremendous increase in the mathematical maturity of the student upon her return. One respondent commented,

> Clearly, the major benefit to [her] was realizing that she was the mathematical equal of some of the most talented women of her age. While her background in course work was not as strong as some, her mathematical training and ability let her participate as an equal. This was exceptionally helpful to her and did wonders for her self-confidence.

Concerning a different student, another wrote,

> [She] mathematically matured a great deal as a result of your program. I noticed that the analytical and topological concepts ... meant a lot more to her than they did to other students.

The students with whom the six respondents had no contact were from Princeton, Stanford, Berkeley, and Johns Hopkins. All of these students did stay in close contact with their SMI professors, receiving letters of recommendation and advice about graduate school. Two of them even wrote undergraduate theses under the long-distance supervision of their SMI professors. We interpret this as evidence of the benefit of the SMI to undergraduate women at major research universities: the SMI provides them with invaluable support that is missing at their home institutions.

The faculty respondents also provided information on the record of their institution in sending women to graduate school in mathematics. At roughly one-quarter of the schools, at most one female student goes to a graduate program in mathematics in any five-year period; at one-quarter, one female student goes every other year; at one-quarter, one goes about every year; and at the remaining quarter, about two go each year. These numbers make it clear that undergraduate women usually have little contact at their home institutions with other women who plan to attend graduate school in mathematics. By way of contrast, at the SMI, students find themselves part of a significant group of women dedicated to math, over two-thirds of whom go on to do graduate work in the mathematical sciences.

Results from the graduate advisor survey.

Seventeen of the 1991-92 SMI students who responded to our survey have begun work on a thesis, and of these, 14 allowed us to contact their thesis advisors. The advisors' responses were unanimous: each student is making satisfactory progress toward her degree. Many advisors responded with accolades such as: "one of our best in the past five years," "the most dedicated student that I have ever seen," "a model graduate student," and "a lot of self-motivation." We see a confirmation of these opinions in the quantitative part of the survey. According to the advisors, the students have a great deal of self-confidence and motivation, and they arrived at graduate school knowing what to expect.

When asked "How does the student compare to other women in your program in terms of adjusting to the demands of graduate work?" the advisors had insightful comments, and offered great encouragement. For example, one advisor wrote:

> As an undergraduate, [she] attended a small state college and received an education that really didn't give her the necessary background for graduate school. Nevertheless, she arrived here in graduate school imbued with confidence and the desire to work hard ... She has passed all her qualifying examinations and is now writing a dissertation under my direction. She is a joy to have as a student; she's talented and energetic. If your program had anything to do with this, then you should certainly consider it a success.

What happens to the faculty that teach in the program?

We asked 12 seminar leaders from 1991 to 1995 (excluding current and future directors) to tell us what effect the program had on them. All responded to our request. They resoundingly confirmed that the SMI was a valuable opportunity for them as well as for the students. Here are excerpts from three statements.

From a 1991 seminar leader:

> [The program] had a big and beneficial influence on me. I gained a unique teaching experience. The experience was beneficial for me from the point of view of my mathematical research also. Collaboration with Steven Givant in the theory of relation algebras turned out to be very fruitful and is marked with several joint papers since then. I had mathematical collaborations with other researchers in the Bay Area (e.g., William Craig, Richard Thompson).
>
> I would recommend working in this program to any good mathematician, because I find it very precious and unique,

and because it is really enjoyable for creative people.

From a 1993 seminar leader:

> I cared about every single one of the women I had in my seminar at the SMI (they were all potential future women professors in the mathematical sciences), so I was highly motivated to make the seminar a wonderful experience for every student. This is precisely the attitude that ... every liberal arts college professor must cultivate. ...
>
> The SMI has created a network of women mathematicians nationwide. This year I have an NSF Visiting Professorships for Women grant. When I think about whom to invite to talk about doing mathematics with women graduate students, ..., I automatically think of professors who have taught in the SMI. My professional ties with several colleagues in my field have been strengthened because we share the experience of teaching in the SMI and related summer programs for women.
>
> I think it is a wonderful service opportunity for women mathematicians with established research reputations.

From a 1994 seminar leader:

> I have always tried to get students to participate actively in the classroom. However, while I was teaching in the SMI, I was able to engage students more both in my presentations of new material and through problems that they worked on in groups. After my experience in the program, I developed a clearer idea of how to ask questions in class which would get the students involved in thinking about the material, and telling me how the proofs should be done. Now almost all the proofs which are presented in my classes are constructed by the students. Also, I began putting deeper problems on my homeworks, and making more of an effort to get students to work in groups. As a result I think that the students are learning the material better and becoming more interested in the material in my classes.

Conclusion

The Mills Summer Mathematics Institute is a highly successful program for mathematically talented undergraduate women that has received national attention. It has demonstrated that an intense mathematics program, which provides strong role models and a network of support, can be a key factor in positively influencing women to pursue graduate studies and to plan careers in mathematics. The program evaluation clearly shows the tremendous and lasting impact of the program on its participants.

Summer Program for Women in Mathematics

MURLI M. GUPTA AND DANIEL H. ULLMAN

The George Washington University

During the summer of 1995, The George Washington University (GW) Mathematics Department hosted a four-week pilot program for 10 outstanding undergraduate women mathematics majors from around the United States. This program was funded by the National Security Agency (NSA). The program was a terrific success. Each of the 10 participants was glowing in support of the program. Every one of the directors, instructors, and teaching assistants felt that the program provided a tremendous benefit to all participants. In 1996, we hosted a similar program for 16 undergraduate women, and the program is continuing for the summer of 1997.

Overview

The Summer Program for Women in Mathematics (SPWM) is designed for outstanding undergraduate women who are majoring in mathematical disciplines, who have completed their junior year, and who are considering graduate study in the mathematical sciences. The goals of the program are to communicate an enthusiasm for mathematics, to develop research skills, to cultivate mathematical self-confidence and independence, and to promote success in graduate school. We bring the participants into contact with successful women mathematicians in academia, industry, and government. We aim to provide the students with a broad exposure to mathematical culture, illustrating the beauty and attraction of mathematics, the tools necessary for success in mathematics, applications of mathematics to business and industry, and the career opportunities available to mathematicians.

The academic program is centered around two three-week courses, starting the first and the second weeks of the four-week program. In each of the first and last weeks, a one-week minicourse features some new areas of mathematical study. All courses are led by professional women mathematicians, who are assisted by graduate students. Topics for these courses aim to complement the typical math major curriculum and are focused to permit the students to reach interesting open problems in a relatively short time. The plan is to lead the students to the forefront of current research, so that they might learn to appreciate the mathematical research enterprise.

Throughout the program, we provide extensive contact with guest speakers who give expository talks on the areas of their research interests. Some of the guest speakers also address mathematical history, mathematical ethics, and mathematical philosophy. The guest speakers are available to participate in discussions about their careers and personal and professional experiences.

In addition, we organize panel discussions on the issues of careers and the job market, graduate schools, and gender. We arrange field trips to see women mathematicians at work in the many centers of mathematical activity in the Washington area. A series of mathematical films is offered, and the program allows ample time for self-paced work as well as for reflection, recreation, and relaxation.

SPWM participants are housed on the GW campus in downtown Washington, DC. In order to foster an atmosphere of community and camaraderie, all student participants and teaching staff (including graduate students and professional mathematicians who lead the instructional program) reside in a dormitory. The participants have access to the library, computer, and recreational facilities on the GW campus.

The Washington area is an ideal location for a program to immerse students into mathematics. There is, around Washington, a thriving community of pure and applied mathematicians working at federal government agencies and laboratories, at seven major universities, and in high-tech industry. The nation's single largest employer of mathematicians, the National Security Agency, is also located near Washington. We take advantage of our unique location by visiting mathematicians at such sites.

Instructional Activities

During each of the four weeks of the summer program, we devote four days to activities based on campus and one day to a field trip associated with careers and applications of mathematics. Weekends are left for independent and group study, consultation with faculty and graduate students, rest, and relaxation. The activities based on campus are centered around two three-week courses, starting the first and the second week; there are also two one-week minicourses, during the first and fourth week of the summer program.

Each session is led by a professional woman mathematician and a teaching assistant. Each course focuses on an accessible area of current research and involves the participants in group work, problem solving, mathematical writing, speaking, library research, and computation.

Each course provides a learning environment in which lecturing plays a minimal role, with the staff members doing more questioning than answering, more guiding than revealing. Teamwork is encouraged.

The students have access to computer labs on campus, providing a flexible environment for symbolic and numerical calculation. They have access to the internet and through it to the worldwide web. They also have access to the Gelman Library at GW, which subscribes to over 100 mathematical journals.

The students take an active role at all times. Working individually and in groups, the students explore, experiment, discover, formulate, conjecture, and prove significant mathematical results. Oral and written presentation are an important component; on the last days of each course, the students present written and oral reports summarizing the course. In this environment, students discover that they have the power to do mathematics on their own. They develop the self-confidence to engage in independent work and the necessary communication skills to engage in the kind of collaborative efforts that produce much of today's new research.

In 1996, Professor Darla Kremer, Murray State University, offered a three-week course entitled "Polya Theory, Symmetric Functions, and Group Representations"; Professor Karma Dajani, University of Utrecht, offered a three-week course entitled "The Dynamics of Expansions"; Professor Janet Kammeyer, US Naval Academy, offered a one-week minicourse entitled "Classification Problems in Ergodic Theory"; and Professor Jaye Talvacchia, Swarthmore College, offered a one-week minicourse entitled "The Classification Problem for Compact Oriented Manifolds: Classical Results and Open Questions."

The professors were assisted by graduate students Natalie Priebe, University of North Carolina, and Elizabeth Wilmer, Harvard University.

In 1995, Professor Leila Schneps, Ecole Normale Supereieure, Paris, offered a three-week course entitled "Groups and Numbers; an Introduction to Classical and Inverse Galois Theory"; Professor Lauren Rose, Wellesley College, offered a three-week course entitled "Algebraic Combinatorics on Convex Polytopes"; and Dr. Desiree Beck, National Security Agency, offered a one-week minicourse entitled "An Exploration of Partitions, Young Tableaux, and Symmetric Functions."

The faculty were assisted by graduate students Cathy O'Neil, Harvard University, and Darla Kremer, University of Iowa.

Guest Lectures

The program of guest speakers is intended to bring the participants into contact with a wide variety of mathematical professionals. We invite two guest speakers each week. We coordinate the topics for the guest talks with the mathematical content of our classroom activities, both by preparing the students beforehand and by allowing time for discussion afterwards. The speakers interact with the participants after their talk and entertain discussions on their background, their education, and their careers. Our participants found that they obtained "more glimpses into the wide, diverse world of math," and that they enjoyed the opportunity to interact with the speakers before and after the lectures.

Following is a list of speakers and titles for Summer 1996:

Michael Moses, George Washington University, "Robertson & Seymour, Kuratawski & Hercules, Hilbert & Godel, and Harvey Friedman... and more!"

Ann Trenk, Wellesley College, "How to pay your faculty fairly using k-leveling functions for posets?"

Karla Hoffman, George Mason University, "Operations research and decision making."

Linda Lesniak, Drew University, "Postman versus salesman- who has the advantage?"

Leslie Hall, Johns Hopkins University, "Combinatorial scheduling: algorithms and anomalies."

Jim Schatz, National Security Agency, "Constructing designs for group actions."

Jane Hawkins, University of North Carolina,

"Lebesgue, ergodic, rational maps of the sphere and where they live in parameter space."

Evenings

Many evenings are focused around problem sets. The problems ask students to explore examples, conjectures, fallacies, paradoxes, definitions, theorems, and generalizations. The faculty and teaching assistants are available during these evening hours to stimulate or encourage student participants and to assist in the development of collaborative problem solving groups.

Friday evenings are devoted to mathematical films, selected on the basis of their mathematical content, the quality of their presentation, and their cultural and philosophical insights. The films supplement our guest lectures and the curricula in our classrooms; each show is followed by a group discussion. The GW library has an extensive collection of appropriate films.

In the past years, we have shown the following films:

- The Mathematical Mystery Tour
- The Birth of Modern Geometry
- The Liberation of Algebra
- The Story of Pi
- Outside in
- Not Knot
- The Vernacular Tradition
- N is a Number

The participants find these films to be entertaining and informative and have gained historical and cultural perspectives that are not customarily present in a typical mathematics classroom.

Field Trips and Panel Discussions

One day each week is devoted to an area of application of mathematics and an associated field trip. These activities bring the participants into contact with women mathematicians in their own workplace and expose the participants to current issues at the forefront of mathematics, the wide variety of applications of mathematics, the depth and complexity of the kinds of mathematics involved, and the possibilities for careers related to mathematics.

In 1996, the field trips were to

- National Institutes of Health, Bethesda, Maryland
- NASA Goddard Space Flight Center, Greenbelt, Maryland
- Smithsonian Institution, Washington, DC
- National Security Agency, Ft. Meade, Maryland

We also take the participants to visit the Maryland Science Center in Baltimore to view the exhibit entitled "Beyond Numbers." This exhibit opened in June 1995, and was designed with the help of four members of the GW Mathematics Department, including Professor Rodica Simion.

We organize several panel discussions to address other issues associated with the mathematics community, including careers, the job market, and graduate schools. Among our guest panelists in 1996 were Leslie Gruis of the National Security Agency, Rachelle Heller of the School of Engineering and Applied Sciences at GW, Allyn Jackson of the American Mathematical Society, and Jennifer Zito of the Center for Computing Science.

Selection of Participants

Our participants are women undergraduates who have completed their junior year and are considering graduate study in mathematics. We are especially interested in attracting students who might not have access to, experience with, or information about graduate study in mathematics. We expect students to have some experience with mathematics beyond the typical first courses in linear algebra and differential equations.

We mail program announcements and application packets to all degree-granting institutions in the United States. We also make program announcements through electronic networks and print journals, and through electronic mailings to previous participants and other individuals. Our home page at

http://www.gwu.edu/~math/spwm.html

is visited by a large number of interested students who can download the application material directly from the program homepage.

The selection process is based upon the students' college transcripts, personal statements, and letters of recommendation.

Program Assessment

We carry out program evaluations through extensive formal and informal feedback from the participants. In 1995 and 1996, we asked the participants to provide written comments at the midpoint and at the end of the program.

Some representative comments were:

"I gained a wealth of information and insight. I'm now convinced that I can succeed in grad school."

"Just being around peers and faculty and other professionals helped a great deal and I've made a [career] decision."

"I feel the strongest part of this program was the interaction between us the students with faculty, guest speakers, and TAs. I have learned so much from it and it has made this experience one which will have a strong impact on my life hereafter."

"It was great reading the research papers. It was very encouraging to realize that I know enough to read other people's work."

"Although we worked really hard, I think it was a good prep for grad school."

"I would say that this has been the best experience of my undergrad career in many ways."

"I am amazed at how much more aware I am now than I was a month ago."

In addition, an advisor for a 1995 program participant reflected in 1997 that

"When she returned to [school] in September, she was unable to contain her enthusiasm. After her summer experience, she was certain that she wanted to pursue a career as a professor of mathematics."

We conclude from these and similar remarks that SPWM is successful and is providing a much needed resource to the nation's undergraduate women. We expect to be able to continue this program at GW for many years to come.

Carleton and St. Olaf Colleges' Summer Mathematics Program[1,2]

DEANNA HAUNSPERGER AND STEPHEN KENNEDY

Carleton College

Several years ago, Amy, a student of ours, came back from the Mills Summer Mathematics Institute bubbling with enthusiasm and ready to begin applying to graduate school in math. We talked to Amy about the experience, and she couldn't say enough good things about the wonderfully inspiring, nurturing environment. She confessed to just one regret: she had learned so much about how to prepare for graduate school, and she only had a year left as an undergraduate—she wished that she could have gone when she was younger.

The next year, when we saw a notice in the AWM newsletter about the formation of a Consortium to Advance Women in Mathematics and a call for proposals for programs for women, we remembered Amy's words. We enlisted the aid of our colleagues Laura Chihara and Kay Smith from St. Olaf and Gail Nelson from Carleton and started planning a program for students finishing their first or second year of college.[3] This, we felt, was a crucial time to catch young women, a time when they are making difficult, perhaps irrevocable, decisions about their futures. And they are making these decisions without a lot of accurate information—every year dozens of young women apply to our program and tell us "I love math, but I really don't want to teach—that's why I'm double-majoring in biology (or chemistry, or business)." During our time at St. Olaf and Carleton we have learned that a word or two of encouragement can make an enormous difference to a first- or second-year student, especially an insecure, unconfident student; and having a critical mass of students together studying math, going to colloquia and socializing builds an incredibly strong sense of community. We thought that we could use some of what we had learned to affect a group of young women and

give them some tools and, more than anything else, the desire and self-confidence to continue in math.

What we do

The heart of the program is, of course, the coursework. The women take two classes which meet 8:30 to noon, Monday through Friday morning (one class meets MWF, the other TuThF). We have been extremely fortunate in our choice of instructors, each of whom has thrown herself fully into the program and given much more than we expected. We always have one instructor from a research university and one from either Carleton or St. Olaf (each aided by a female teaching assistant from St. Olaf or Carleton). In the first year, Judy Kennedy of the University of Delaware led a Moore-method seminar in point-set topology and a computer-experiment-driven course in dynamical systems. These were woven together at the end when she explained how the strange and beautiful, almost Baroque, abstract point-set theory becomes useful when trying to understand the strange and beautiful unexpected behavior of some dynamical systems; this connection, of course, is the focus of most of her research program of the past decade. The students were stretched and challenged in a way they could never have been in a regular college class—and they rose to the challenge. By the end of the program 17 of the 18 women had solved and presented to the class their solution to at least one of the problems on The List. The rigors of Judy's class were well-complemented by a more leisurely tour through knot theory led by Gail Nelson of Carleton.

In the second year the demands of the two classes were more balanced as Tami Olson of Michigan Tech taught applied functional analysis and Laura Chihara of St. Olaf taught algebraic coding theory. In both years, the students reported in post-program evaluations that they had been asked to do things they they did not believe they could do, but with the support of the faculty and of their peers, they had struggled to accomplish something difficult and meaningful. This year Laura Chihara will repeat her coding theory course, and Karen Brucks of the University of Wisconsin at Milwaukee will teach a course in low-dimensional dynamical systems.

We keep the students very busy—in addition to the coursework there is a twice-weekly colloquium series. Highlights have included Mary Ellen Rudin's visit (A History of Women in Mathematics), an introduction to Clifford Algebras by Marge Murray of Virginia

---

[1] See our web page at:
http://www.mathcs.carleton.edu/smp.

[2] The directors would like to express their great appreciation to the National Science Foundation and the National Security Agency for funding.

[3] The current directors are Laura Chihara, Deanna Haunsperger, Stephen Kennedy, and Gail Nelson.

Tech, and our annual visit from Joe Gallian (University of Minnesota, Duluth) who not only gives a lovely colloquium, but also tells the students about the opportunities available to them in REUs. We always have talks by several local faculty, and we try to bring in a speaker or two from outside academia (each year we have invited a speaker from the National Security Agency and last year an epidemiologist from the Minnesota Department of Health). We do usually have one or two male speakers, but we try to have a preponderance of females.

In addition to the colloquium series, there are weekly panel discussions. The topics are: completing a math major; careers that use mathematics; applying to and succeeding in graduate school; and women in science. One afternoon a week the students have a three-hour session in a computer lab learning Mathematica, LaTex, html, or Java. One evening per week we have recreational problem-solving.

Around and through it all, we weave mathematical outings and social activities: a tour of the Minnesota Geometry Center, at least one picnic per week, movie nights, canoe trips, hikes, even the Mall of America. Every weekend features at least one organized excursion. One night per week Deanna visits the dorm lounge for "Deanna chat," a time for her and the students to check that everything is going smoothly. The program closes with a banquet celebrating success and honoring the participants; farewells are difficult for the women, many of whom found long-sought-after sisters in mathematics.

The Outcome

Our 18 young women mathematicians immerse themselves in mathematics, living and working in a supportive community of women scholars (undergraduates, graduates, and faculty) who are passionate about learning and doing mathematics. Our intentions for them are threefold: to excite them about mathematics and mathematical careers; to provide them with some of the tools they will need to succeed in a mathematical career; and to connect them to a network of fellow female mathematicians. The 1995 and 1996 Carleton-St. Olaf Summer Mathematics Programs (SMP), as measured by the participants' post-program evaluations, were successful in achieving all of these goals.

The students, faculty, teaching assistants, and directors all confess to being profoundly influenced by the program. Every faculty member has reported that this was the most rewarding teaching experience of her life. The following quotes are culled from post-program evaluations:

> This experience has revived my mathematical soul and charged me up.

> Thank you for an awesome experience. It is something I will remember for the rest of my life.

> This has been, by far, the most exciting and **fun** experience in math I've ever had.

> The program has certainly confirmed my desire to major in math at the undergraduate level, and it has revealed more options than I previously expected for graduate study and an eventual career.

> But the more I listen and learn of myself, I know that I can't stop after college and I won't stop until I feel that I've learned enough. And when I come to your grad school panel in a few years I'm gonna tell some really good stories.

> To learn about other womens' experiences, to be encouraged, supported, to have so many people believe in me, and to connect with such brilliant and fun women was awesome!

> I found it inspiring and informative to be around so many knowledgeable people with whom I could easily connect and communicate. The information and contacts that I have gained are invaluable. Everyone should have something like this at some point (sooner rather than later).

The students return to their home institutions eager to plunge into their studies. They have a clearer idea of what mathematics is and how to organize their future plans. Their increased awareness of various topics within mathematics have led many to give talks in their home departments on the mathematics that they learned in the summer program. Most have already, or intend to, participate in REUs, the Budapest Semester, or other enrichment programs.

All who have done so acknowledge being much better prepared to succeed at, and benefit from, those programs than they otherwise would have been. Perhaps more important than the knowledge and renewed excitement for mathematics, each of the students has gained confidence in her ability to do mathematics.

> It challenged me, but I was able to work through the proof, I really put my heart into it—and I loved it. It was not only good for me alone, but it was so special to have a class full of people who could handle this level of self-motivated, rigorous learning.

> But now, given this opportunity, I'm excited for school to start in the fall and I'm excited that I am a smart and intelligent math student. Really I am. And most of all, I don't need to prove it to anyone—just to myself.

> [Studying mathematics in a group of women students] is the best! People explained things so that others would understand, and people kept telling each other, "Good call," "great idea," or "you're brilliant." You don't hear that studying with guys. It is very reassuring to discover that almost everyone else has the same insecurities and self-doubts and when you realize everyone else's are unfounded, it starts to chip away at your own. A very positive experience.

> Most of all the program has given me the confidence that I can succeed in math, both as a student and as a woman.

> I have proved things which I had doubted I could.

> The satisfaction one derived from finally completing that proof they had been working on for a week was tremendous. It taught you that you could do things on your own.

> The [SMP] has given me the confidence, the mathematical foundation, and the desire to propel myself through advanced studies of mathematics. It has been

an immense force driving all of us on to higher plains of mathematics. The courses and people contributed to an environment which has nourished and developed our mathematical souls.

This confidence building is central to the mission of this program. All of these students, and most of the other 100 who applied, are intellectually capable of achieving an advanced degree in mathematics. Something other than intellectual capacity prevents many women from pursuing one. Heightened self-confidence and a supportive network of colleagues and mentors are two factors which we hope prevent young women from dropping out. These students return to their home institutions knowing that women can and should be doing mathematics. They will not only be supported by this knowledge, but they also will carry the message back with them to influence their peers and their teachers.

Who are the participants?

The program is advertised each year through a mass-mailing to all bachelor's and advanced degree-granting mathematics departments throughout the country. Also, advertisements are placed in the AWM Newsletter, FOCUS, Math Horizons, and several electronic newsletters. Participants have been first- and second-year undergraduates from regional and national liberal arts colleges and universities throughout the country. The "typical" student has had three semesters of Calculus, a Linear Algebra course, and one to three other math courses. She has a GPA of 3.85, she works in the campus tutoring service, and she plays the French horn. The directors attempt to choose the participants so that their academic backgrounds are somewhat balanced—we have, however, on occasion admitted women whose preparation didn't match up but whose apparent talent was such that we thought they could handle whatever we threw at them. We have also erred—each year there have been a couple of students who see the passion and commitment of their peers for mathematics and decide that mathematics is not that important to them.

Conclusions

One of the things that most surprised us about these young women is that, by and large, they themselves do not believe that their gender has had an impact on their education. In fact, many adamantly maintain that they have never been the victim of gender-based inequity. And yet, in their applications, many students do mention a desire to study mathematics with other women, and many applicants do notice the gender imbalance in their math and physics classes. At the close of the program, they maintain that studying in this all-female group was the most supportive, enriching, and joyous intellectual experience of their lives. They don't, or many don't, ask themselves why all of their schooling could not be this nurturing and exciting. Of course, there are other factors at work besides the all-female environment, e.g., the intensity and richness of the program which would be difficult to duplicate in an ordinary classroom. Nonetheless, we believe that many of these young women speak out and take chances here in a way that would not happen in a coed environment.

Every year we are forced to reject over 100 young women—most of whom would benefit from an experience like this. There are many fabulously talented young women at small two- or four-year colleges whose chances to obtain an advanced mathematics degree would be greatly enhanced by an experience like what we offer. We see room for several more programs, each aimed at a slightly different group of these students. In particular, we have carved out a sort of middle level of necessary preparation: we choose mostly students who have only one or two theoretical courses after Linear Algebra. It would be easy to fill up two more programs like ours: one for women who have come to mathematics too slowly and have yet to get beyond Linear Algebra; and another for women who approached mathematics too greedily and have too much experience in upper-level classes to fit in well here.

The young women who apply to our program seem to be very concerned about getting a leg up on establishing a career—most don't know what kind of career, but they are worrying about it already. It seems to us this might in part explain why so many of them, while clearly possessing the ability and love of mathematics that would ensure success in graduate school, choose not to pursue advanced degrees. Perhaps young men in college, for whom the existence of a career seems virtually assured in our society, feel they can afford the luxury of indulging their intellectual curiosity more than young women do?

It is clear to us that we have an impact on the lives of the young women who come to our program—we see the increased confidence, enthusiasm, knowledge, and mathematical sophistication. We see the electronic messages they post on the program's listserver to let us and each other know what is happening in their lives—mathematical and otherwise. It is less clear to us how to measure this effect. We can never know how many would have gone on to productive mathematical careers without us—given the talent level, some certainly would. We won't know for some time how long and how far the impetus we give sustains them in the face of adversity. In any case, we are operating with very small numbers; the few programs like ours will not by themselves even out the gender imbalance in our profession. We do believe that we are making a difference though: just today, Suzanne (SMP '95) sent e-mail to tell us that she had been accepted for graduate study in mathematics at Berkeley and that she was considering accepting; she felt strong enough and smart enough. Go Suzanne!

The Mount Holyoke Summer Research Institute

Margaret Robinson

Mount Holyoke College

Over the past quarter century, Mount Holyoke has accumulated much experience with programs that guide undergraduates in summer mathematics research. Our research institute has operated under funding from a variety of sources, including the National Science Foundation, the Pew Foundation, the Howard Hughes Medical Foundation, and Mount Holyoke College. It is based on the belief that students become better prepared for graduate programs by active experience in research. We believe that to build confidence, self-esteem, and mathematical maturity it is extremely important for our students to experience mathematics as a creative process. It comes as a wonderful surprise to undergraduates that they learn large pieces of mathematics while pursuing a good research problem (e.g., in investigating the possible combinations of local maxima, minima, and saddles for a polynomial in two variables or in investigating symmetry breaking in crystal growth).

Each summer we direct the work of 10 to 20 students for approximately nine weeks. The students work in groups of five. Each group is led by a single faculty member and attacks a mainstream but well-focused research problem. Typically, participants are students who have just completed their junior year. However, we do admit some students following their sophomore year, if they are well enough prepared. Each group is assigned a large room equipped with a conference table, a blackboard, desks, a library of relevant texts and papers, workstations, and PCs. In addition, other rooms are available for study and quiet. The groups share a common room with refrigerator, microwave and coffee-maker.

Each group meets twice daily with the faculty. They begin the morning by meeting with the advisor to plan the day's activities, and they meet again in the late afternoon to take stock. The faculty advisor is also in touch throughout the day. The day ends with afternoon tea in the common room. The summer is very intense for both faculty and students and demands a more than full-time commitment from each faculty member. Since the project is of research interest to the faculty members, students are aware that they are working on problems which are difficult and substantial so that their morale remains high even when results come slowly.

Typically, there is much collaboration, and everyone becomes involved in the teaching process. The faculty start the students right off with a beginning, perhaps computational, problem. There are introductory lectures, but the students participate by investigating and presenting pieces of the material themselves. As soon as the problems are understood by the students, the students' own questions and confusions determine the topics of further faculty lectures. As the summer progresses, faculty lecture less, students more.

Young and inexperienced students ordinarily do not have the confidence and expertise to allow them to deal with general, open-ended questions. In order to build confidence and satisfaction, the first three weeks or so are given over to straightforward tasks and questions which have clear-cut answers. Students gradually become capable, as they mature in the program, of dealing with less definiteness and less certainty.

Thursday afternoons are reserved for the program's colloquium, where formal progress reports are given by each group with everyone presenting a piece of the problem. The inexperience of undergraduates is never more painfully apparent than in their first presentations, and we have learned over the years how valuable it is for them to have repeated opportunities to say what they are doing, and, in the process, to clarify what their problem is about and how they are approaching it. This presentation process often results in each student's becoming expert in a different part of the problem.

Frequently, at least one student grasps the problem quickly and starts working on it independently, while, at the other end of the spectrum, at least one student is not in an early position to make much of a contribution and needs faculty attention to stay occupied. A good strategy is to assign a substantial presentation to the weaker student, whereby that student has to learn a good deal about the problem from the quicker students, and perhaps, by this process, become a more independent contributor.

The final two weeks of summer are spent in assimilating, summarizing, and writing. Beginning in the sixth or seventh week of the program, each student is responsible for summarizing her personal progress, and that of her group, in writing, on a weekly basis. Every student takes home a booklet containing the written work of all groups. The writing is a difficult part of the summer experience, but is necessary to seal its success. Past experience shows that students occasionally continue their work in honors theses during their senior year in college.

There is also a weekly lecture series given by distinguished faculty from the New England mathematical sciences community.

In our experience, a good research project should meet the following criteria: It should admit extensive computer experimentation and/or computation. It should deal with mainstream mathematics or statistics. It should have a reasonable expectation of producing results which will be of interest to the larger mathematical community. It should be accessible to able undergraduate juniors with strong backgrounds in mathematics. It should involve some questions which can be almost certainly answered in the allotted time, as well as others which have a promising, but indefinite and open-ended aspect.

Here are descriptions of some of our projects. The unpublished papers listed below are available on the Mount Holyoke College mathematics REU web site at http://www.mtholyoke.edu/acad/math.reu.

A. Effective Methods in Analytic Geometry (directed by Donal O'Shea and Alan Durfee)

The problem is to determine what parts of the Le-Teissier theory of limiting tangent hyperplanes carry over to real algebraic sets. Virtually everything that one might naively expect to hold in the real situation in fact does not, and it is far from clear how to formulate, much less prove, real analogues of the complex results. Progress in settling the problem would have significant practical consequences. The group investigated the algebraic and geometric tangent cones and the limit of normals for certain classes of real and complex surface singularities in three-space. In particular, they looked at a family of singularities given by Trotman:

$$z^b = \pm x^A y^T \pm x^{A+C}.$$

They calculated the geometric tangent cone (the limit of secants) for all real surfaces in this family (with $\pm$ just $+$) by brute force. To find the limit of normals is a more difficult problem. They used a general result of Le and Teissier in the complex case, and almost succeeded in proving it in the real case. They found some general results, and also computed the limit of normals for the members of the

Trotman family again by brute force. In addition, they worked out two more algebraic methods using Groebner bases. When the surface had a nonisolated singularity, they also investigated the relationship between the generic singularity and the special one. For plane curves, they completely solved the problem of relating the algebraic and geometric tangent cone; the solution to this problem is probably known to specialists, but had never been written down carefully. The group produced a thirty page paper in addition to the pages of computations of special cases. This paper will be revised by Durfee and O'Shea and will result in one or two publications.

Papers:

1. Banach, D., Billey, S., Campbell, T., O'Shea, D., and Teleman, C. (1991). Limits of Tangent Spaces to Singular Points of Algebraic Surfaces. To appear in Proceedings of the Trieste Singularities Conference 1991. (O'Shea REU 1988).

2. O'Shea, D., and Teleman, C. (1990). Limiting Tangent Spaces and a Criterion for μ-Constancy of Hypersurface Singularities. To appear in Proceedings of Hawaii-Provence Singularities Conference 1990, SLN. (O'Shea REU 1988).

3. Chinta, G., Koelling, M., Lucas, A., Mandell, J., St. Pierre, R., and Zhang, J. The geometric tangent cone and the limit of normals to real surface singularities. Preprint to be revised and submitted. (Durfee and O'Shea REU 1994).

Finally, we mention that the book "Ideals, Varieties and Algorithms" by David Cox, John Little, and Donal O'Shea was inspired by the summer projects that O'Shea had directed, as is a sequel to this book now in preparation.

B. Singular viscous fingering (directed by Mark Peterson)

Planar viscous fingering is closely related to other physical problems such as diffusion-limited aggregation, and has appeared on covers of mathematics and science magazines, because it produces shapes suggestive of fractals. A nice way to formulate the problem is as the time dependence of the conformal map that takes the exterior of the unit disk onto the exterior of a planar domain D. If the only singularities are poles, then the dynamics can be considered to be the motion of these poles. What in fact happens is that some pole hits the unit circle in finite time, producing a cusp in the boundary of D, where the Green's function becomes singular and the problem breaks down. The equations of motion can be rescaled so that the singular problem makes sense, and the dynamics can then be interpreted as the motion of singularities on the unit circle right from the start. The domains are slit domains, and the conformal maps are Schwarz-Christoffel maps. Most remarkably, the dynamics allow the singularities of the map to split, corresponding to the growth of new branches of the slit domains. This singular case, called singular viscous fingering, was discovered by Peterson's REU group in 1988.

Singular viscous fingering describes the time evolution of a conformal map related to a growing domain. The description we have of it, as an infinite system of coupled differential equations for the coefficients in a power series representation of the map, is not very practical, however. In 1992, the REU group made progress in putting the evolution law into a practical form by proving that the locations of certain singularities of the map (they happen to be on the unit circle) move according to an ordinary differential equation conjectured by the previous REU group. This system of ODEs describes only one class of singularities, however. The map has a second class of singularities which also move, and for whose motion we have found no efficient description.

Papers:

1. Ferry, J., and Peterson, M. (1989). Spontaneous Symmetry Breaking in Needle Crystal Growth. *Phys. Rev. A* **39**: 2740-2742. (Peterson REU 1988).

2. Peterson, M. (1989). Nonuniqueness in Singular Viscous Fingering. *Phys. Rev. Lett.* **62**: 284-287. (Peterson REU 1988).

C. Local Zeta Functions (directed by Margaret Robinson)

Here we proposed to have students investigate the form of the Igusa local zeta function for various classes of polynomials. An immediate goal was to duplicate, contrast, and explore the limitations and advantages of different techniques for the calculation of these rational functions. A more difficult problem

was to look for algorithmic methods for these computations. Our distant goal was to prove the rationality of this zeta function in a new way. The students were introduced to p-adic analysis, Poincaré series, and zeta functions. In our 1992 REU program, the group computed the local zeta function for several curves, concentrating and completely understanding the curve $x^n + y^m$ through direct computation as well as through other methods. The group did not succeed in formulating an algorithm for irreducible curves, but was getting close near the end of two months of work. (We received a preprint two months after the REU from C.Y. Lin in Taiwan on such an algorithm, and his ideas go further than the group went that summer.) In the 1995 REU, the group extended the results of the 1992 REU and of C.Y.Lin to reducible curves, and formulated an algorithm in some special cases. They also considered the local zeta function for hypersurfaces with very bad reduction modulo p. One student from the group continued his summer work by writing his honors thesis in this area.

Papers:

1. Field, R., Gargeya, V., Robinson, M., Schoenberg, F., and Scott, R. The Igusa local zeta function for $x^n + y^m$. Preprint. (Robinson REU 1992).

2. Gray, S., and Karlof, K. Igusa's Stationary Phase applied to products of diagonal polynomials. Preprint. (Robinson REU 1995).

3. Grus, J., and Reunman, D. The Igusa local zeta function for Fermat hypersurfaces with exponent p^l. Preprint. (Robinson REU 1995).

4. Reuman, D. (1996). The Igusa local zeta function, Rationality, and the Poincaré Series. Harvard Undergraduate Thesis.

5. Stavenick, R. Computing the Igusa local zeta function for $f(x) = x^3 - xy + y^3$ using Hironaka's resolution of singularities. Preprint. (Robinson REU 1995).

D. Critical points of real polynomials (directed by Alan Durfee)

The 1989 REU project was to investigate the following question: If $f(x, y)$ is a real polynomial of degree d with isolated nondegenerate critical points, what possible combinations of maxima, minima and saddles can occur? The question was suggested by V. I. Arnold to Durfee as being especially suited to undergraduate investigation. The 1989 group produced Durfee et al (1993), which includes several substantial results. The 1992 group continued work on this project. Notable achievements were: Robertson (1992) discovered polynomials with an arbitrary number of local maxima and no other critical points. V. I. Arnold in his capacity as editor of the Journal of Algebraic Geometry has solicited Robertson's paper for publication in that journal. Robertson, using ideas of fellow REU student Feng, also wrote a routine in Mathematica for finding critical points, since the built-in "solve" routine in Mathematica is too computationally expensive and unreliable. He also implemented a new algorithm developed by Paul Pedersen at Cornell for finding these intersections; Pedersen had developed the algorithm for his PhD thesis but had not coded it. Feng, in addition to making useful suggestions to the others, proved some nice theoretical results. Fabbri and Rolloff gave reports on papers and initiated a graphical study of critical points at infinity.

Papers:

1. Durfee, A., Kronenfeld, N., Munson, H., Roy, J., and Westby, I. (1993). Counting Critical Points of Real Polynomials in Two Variables. *Amer. Math. Monthly* **100**: 255–271. (Durfee REU 1989).

2. Robertson, I. A polynomial with n maxima and no other critical points. Preprint solicited by the *Journal of Algebraic Geometry*. (Durfee REU 1992).

E. Topics in comparative number theory (directed by Giuliana Davidoff)

The project examined some old and some new questions about distributions of primes among residue classes with a given difference k. Define a function $\pi(x, k, a)$ which counts prime numbers in the arithmetic progression $a + kd$ by

$$\pi(x, k, a) = \# \quad \{p \leq x | p \text{ is a prime and } \\ p = a + kd \text{ for some } d \in Z\}.$$

In the first and most widely recognized result in the study of finer comparisons of primes among progressions, Littlewood proved that the difference $\pi(x, 4, 3) - \pi(x, 4, 1)$ changes sign infinitely often as x becomes large. About 50 years later, Knapowski and Turan asked whether this same oscillatory behavior exists for any residue classes a and b for a given arbitrary modulus k. They were able to answer many questions, but the general result remained unknown, even for small k. The research project focused on further investigations of the generalization introduced by Knapowski and Turan and of our own, apparently new, one, which asked about the oscillatory behavior of another difference function. Using further work of Bays, Hudson, Spira, and Stark on the first generalization, along with new calculations of zeroes of L-functions done recently by Rumely and our own numerical work, we were able to prove that this new difference changes sign infinitely often in many cases. Extending the work of this summer research project, Peter Sarnak and his thesis student, Michael Rubenstein, have recently settled Knapowski and Turan's question in proving that these difference functions change sign infinitely often.

In addition, two theses emerged from the 1993 REU group, an undergraduate honors thesis by Yi Wang at Bryn Mawr and a Master's thesis by Caroline Osowski at Mount Holyoke.

Paper:

1. Davidoff, G., Osowski, C., Wang, Y., van den Eynden, J., and Winkle, N. (1993). Some results in comparative number theory. Preprint. (Davidoff REU 1993).

The Director's Summer Program at the National Security Agency –Cryptologic Mathematics for Exceptional Undergraduate Mathematicians[1]

Victoria L. Yates

National Security Agency

The Director's Summer Program is the National Security Agency's premier outreach effort to the very best undergraduate mathematics majors in the country. Each summer we invite 25 exceptional students to participate in a twelve week program where they work directly with NSA mathematicians on mission–critical problems. The program is highly comptetitive and is intended primarily for students between their junior and senior year, but exceptional freshman and sophomores are also considered. Graduating seniors are considered too, but they must be enrolled in a mathematics graduate program for the fall.

The goals of the Director's Summer Program are to:

- Introduce the future leaders of the US mathematics community to the Agency's mission and share with them the excitement of working on mathematics problems of national importance,

- Provide a deep understanding of the vital role that mathematics plays in enabling the Agency to tackle a diverse set of technical challenges,

- Encourage bright undergraduate mathematics majors to continue their study of mathematics and pursue careers in the mathematical sciences, and

- Provide solutions to current operational problems.

The students participating in the program work on a broad range of problems involving applications of Abstract Algebra, Geometry, Number Theory, Combinatorics, Graph Theory, Probability, Statistics, and Analysis. For the first two weeks of the summer, lectures on modern cryptologic mathematics are given. After the lectures, the students are presented with about 10 current problems and choose one or two as the focus for their research. All research is documented in a series of papers written by the students near the end of the summer.

Throughout the summer, students develop mathematical theory, apply what they learn to obtain real-time solutions, and experience the excitement of success built on hard work and innovation. Most students find the work at NSA very exciting and challenging, and many decide to return for another summer. State of the art computing resources are available to all students. For the most part programming is done in C in a UNIX environment. MATHEMATICA is available, as is the computational algebra package MAGMA and a variety of statistics packages.

Information about the Director's Summer Program is sent to over 300 colleges and universities across the United States each year. In addition, students who have scored well in the annual William Lowell Putnam Mathematical Competition are invited to apply. Because of the lengthy security processing required, the deadline for applications is 15 October each year. To apply, students simply complete an application or send a resume. At least two letters of recommendation from faculty members familiar with their work, and a copy of transcripts through the current academic year are also required. Students must be US citizens.

---

[1] More information on this program can be obtained by sending e-mail to vicky@afterlife.ncsc.mil

Appendix
Women in Mathematics: Scaling the Heights and Beyond

Conference Program

Day 1

Welcoming Remarks
Lenore BLUM, *Deputy Director M.S.R.I.*

Excellence in Mathematics
Deborah HAIMO, Past President, MAA, *University of Missouri, St. Louis*

The Mills Summer Mathematics Institute and the Multi-Campus Plan
Steven GIVANT, *Mills College*

Report from Carleton/St. Olaf Colleges
Deanna HAUNSPERGER, *Carleton College*

Report from S.U.N.Y. Stony Brook
Matilde TELEMAN, *S.U.N.Y. Stony Brook*

Preparing undergraduate women for graduate school and beyond
Carol WOOD, *Wesleyan University*

Report from the George Washington University
Daniel ULLMAN, *George Washington University*

Report from the University of Chicago
Peter MAY, *University of Chicago*

Geometry and algebra of a regular 4-dimensional polytope
Lynne BUTLER, *Haverford College*

Day 2

Report from Mt. Holyoke College
Margaret ROBINSON, *Mt. Holyoke College*

Report from the University of Michigan
Donald J. LEWIS, *University of Michigan*

Report from Spelman College
Wanda PATTERSON, *Spelman College*

Reactions to Carleton/St. Olaf and SUNY Stony Brook programs
Wei-Jin HARRISON, *American River College*
Agnes TUSKA, *C.S.U. Fresno*

How can faculty motivate and prepare students for graduate school?
Lidia FILUS, *Northeastern Illinois University*

How do we design upper-level courses to attract women and help them excel?
Lynne BUTLER, *Haverford College*
Michael COLVIN, *Cal Polytechnic*

Do we treat our undergraduate women differently than our undergraduate men?
Chris LEARY, *S.U.N.Y. Geneseo*
Deborah NOLAN, *U.C. Berkeley*

Reactions to the George Washington University and Mills College programs
Magnhild LIEN, *C.S.U. Northridge*

Day 3

Report from the N.S.A. Summer Directorship Program
Barbara HAACK, *National Security Agency*

Reactions to the University of Michigan and Mt. Holyoke College programs
Iwona GRZEGORCZYK, *U. Mass. Dartmouth*
Maura MAST, *U. Northern Iowa*

Reactions to the University of Chicago program
Adam HAUSKNECHT, *U. Mass. Dartmouth*

Reactions to the multi-campus proposal
Lynne BUTLER, *Haverford College*

Closing Remarks
Leon HENKIN, *U.C. Berkeley and Mills College*

Participant List

Ani Adhikari
University of California, Berkeley CA

Antonia Bluher
National Security Agency, La Jolla CA

Lenore Blum
Mathematical Sciences Research Institute,
Berkeley CA

Lynne Butler
Haverford College, Haverford PA

Mike Colvin
California Polytech State University,
San Luis Obispo CA

Nancy Cunningham
Rice University, Houston TX

Lloyd Douglas
National Science Foundation, Arlington VA

Lidia Filus
Northeastern Illinois University, Chicago IL

Erica Flappan
Pomona College, Claremont CA

Steven Givant
Mills College, Oakland CA

Concha Gomez
University of California, Berkeley CA

Iwona Grzegorczyk
University of Massachusetts, Dartmouth,
North Dartmouth MA

Barbara Haack
National Security Agency, La Jolla CA

Deborah Haimo
University of Missouri, St. Louis MO

Wei-Jen Harrison
American River College, Sacramento CA

Gary Hart
California State University, Dominquez Hills CA

Deanna Haunsperger
Carleton College, Northfield MN

Adam Hausknecht
University of Massachusetts, Dartmouth,
North Dartmouth MA

Leon Henkin
Mills College, Oakland CA and
University of California, Berkeley CA

Lily Khadjavi
University of California, Berkeley CA

Stephen Kennedy
Carleton College, Northfield MN

Chris Leary
State University of New York, Geneseo NY

Jim Lepowsky
Rutgers University, New Brunswick NJ

Don Lewis
University of Michigan, Ann Arbor MI

Magnhild Lien
California State University, Northridge CA

Miloufer Mackey
State University of New York, Buffalo NY

Maura Mast
University of Northern Iowa, Cedar Fall IA

Peter May
University of Chicago, Chicago IL

Julie Mitchell
University of California, Berkeley CA

Deborah Nolan
University of California, Berkeley CA

Colette Patt
Physical Sciences Affirmative Action Officer,
University of California, Berkeley CA

Wanda Patterson
Spelman College, College Park GA

Ami Radunskaya
Rice University, Houston TX

Brooks Reid
California State University, San Marcos CA

Margaret Robinson
Mount Holyoke College, South Hadley MA

Matilde Teleman
State University of New York, Stony Brook NY

Roy Thomas
Professional Development Program,
University of California, Berkeley CA

Agnes Tuska
California State University, Fresno CA

Daniel Ullman
George Washington University, Washington DC

Sunita Vatuk
Princeton University, Princeton NY

Monica Vazirani
University of California, Berkeley CA

Carol Wood
Wesleyan University, Middletown CT

Susan Yeh
Mills College, Oakland CA